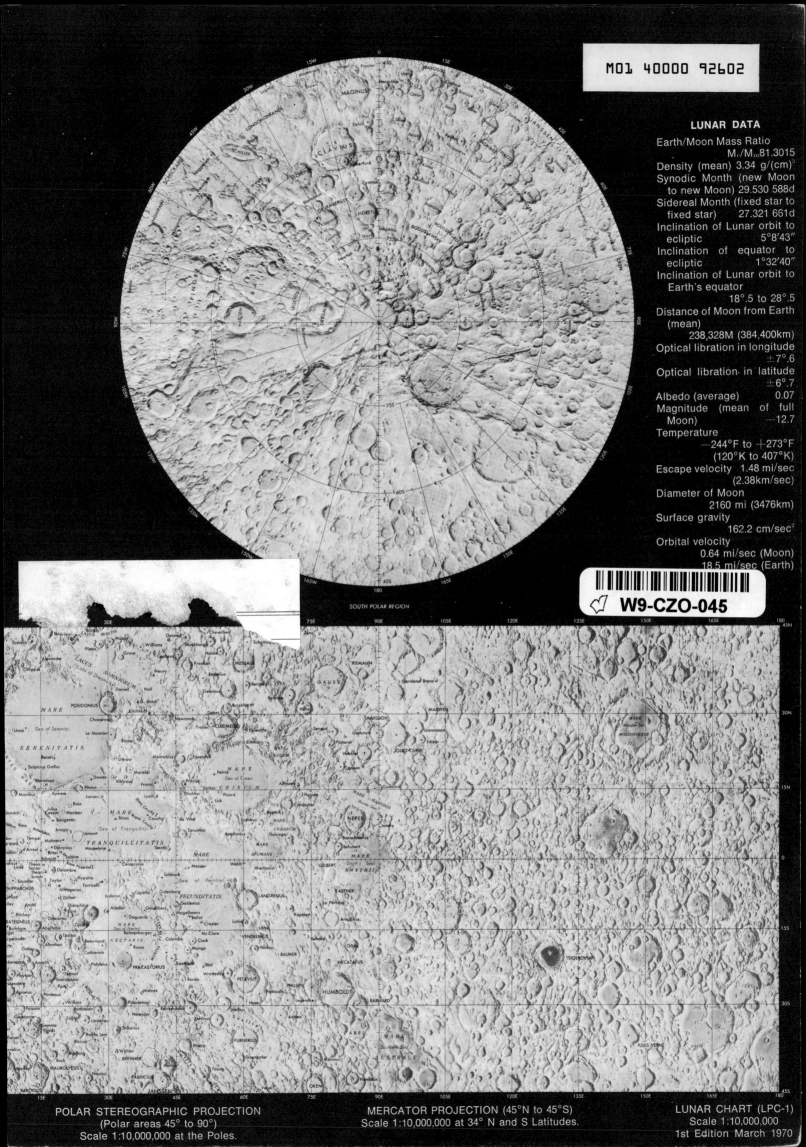

MO1 40000 92602

LUNAR DATA

Earth/Moon Mass Ratio
M_e/M_m 81.3015
Density (mean) 3.34 g/(cm)³
Synodic Month (new Moon
to new Moon) 29.530 588d
Sidereal Month (fixed star to
fixed star) 27.321 661d
Inclination of Lunar orbit to
ecliptic 5°8′43″
Inclination of equator to
ecliptic 1°32′40″
Inclination of Lunar orbit to
Earth's equator
18°.5 to 28°.5
Distance of Moon from Earth
(mean)
238,328M (384,400km)
Optical libration in longitude
±7°.6
Optical libration in latitude
±6°.7
Albedo (average) 0.07
Magnitude (mean of full
Moon) −12.7
Temperature
−244°F to +273°F
(120°K to 407°K)
Escape velocity 1.48 mi/sec
(2.38km/sec)
Diameter of Moon
2160 mi (3476km)
Surface gravity
162.2 cm/sec²
Orbital velocity
0.64 mi/sec (Moon)
18.5 mi/sec (Earth)

W9-CZO-045

SOUTH POLAR REGION

POLAR STEREOGRAPHIC PROJECTION
(Polar areas 45° to 90°)
Scale 1:10,000,000 at the Poles.

MERCATOR PROJECTION (45°N to 45°S)
Scale 1:10,000,000 at 34° N and S Latitudes.

LUNAR CHART (LPC-1)
Scale 1:10,000,000
1st Edition March 1970

A NEW PHOTOGRAPHIC ATLAS OF THE MOON

Other books by the same author

CLOSE BINARY SYSTEMS (1959)
NUMERICAL ANALYSIS (2nd ed., 1961)
EXPLORATION OF THE MOON BY SPACECRAFT (1968)
TELESCOPES IN SPACE (1968)
THE MOON (2nd ed., 1969)
WIDENING HORIZONS (1970)

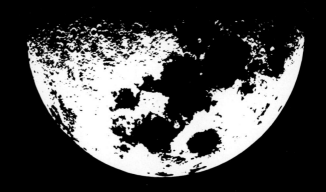

A NEW PHOTOGRAPHIC ATLAS

OF THE

MOON

ZDENĚK KOPAL

INTRODUCTION BY HAROLD C. UREY

First Published in the United States in 1971 by Taplinger Publishing Co., Inc., New York, New York

ISBN 0-8008-5515-9

Library of Congress Catalog Card Number 72-125480

Published simultaneously in the Dominion of Canada, by Burns & MacEachern Ltd., Ontario

Designed by Charles Kaplan

Printed in the United States of America

Foreword

The Earth has been studied intensively for the last one and a half centuries, and a most fascinating history of our planet has been established by the patient and often very physically laborious studies by geologists of the Earth's surface. These studies have covered extensive volcanic processes, the complicated effects of running water, the deposits in the quiet water of oceans and lakes and the presence of the fossil remains of living organisms. This history has been carefully dated in the years of this century by radioactive dating methods. It is especially well established for the last 600 million years since the beginning of the Cambrian period, but considerable information exists for times preceding this. However, the oldest rocks so far found are about 3 to 3.5 billion years old, and at present it appears that the study of the Earth's surface will give us very little in regard to any earlier history. The Earth has been subjected repeatedly to the building of great mountain chains and their subsequent erosion by water.

A brief and cursory glance at other planets which may be investigated by the space program is not very encouraging. The photographs of the surface of Mars indicate that it has been subjected to erosion, though probably not as extensively as on the Earth, but yet of such magnitude that its history probably will be difficult to interpret. Venus is too hot to investigate in great detail, and probably has had as intense a volcanic history as has the Earth. Mercury will be difficult to investigate. The giant planets have left a very obscure record of their historic past. Thus, only the meteorites and the Moon present us with some durable samples which have preserved the record for the time before 3 or 3.5 billion years ago.

The ages of many meteorites of various kinds lie near 4.6 to 4.7 billion years with a curious exception of one iron meteorite with a silicate inclusion which was last melted about 3.8 billion years ago. The surface of the Moon, with its intense collisional record, must have acquired this collisional history before the oldest rocks of the Earth were laid down. Otherwise, these terrestrial rocks would show a similar record of such a history which is not the case. Indeed, the dating of these lunar rocks show that the lunar surface acquired its general chemical composition by some melting process about 4.6 billion years ago, and that some rocks were last melted about one billion years later. Indeed, the Moon is a very old object having existed as a separate planetary object from a time near the beginning of the solar system, as indicated by the meteorites. Though some volcanic activity certainly has occurred, much of the Moon's surface must preserve the record of a very ancient history. However, at the present time, there is no evidence pointing to an earlier history than that of the planets.

The present volume presents a collection of some of the very best pictures of the lunar surface which have been secured from Earth-based apparatus and from cameras on space vehicles near the Moon. The former are limited to the near hemisphere facing the Earth, but the latter include excellent pictures of the far hemisphere and much more detailed pictures of all areas of the Moon including the polar regions and the areas of the limb which could be secured only very imperfectly by terrestrially based telescopes.

As is evident to all students of the Moon, it is not possible from pictures alone to understand the physical processes that have shaped the Moon's surface. Thus, most students have believed, from the photographic evidence, that both collisional and volcanic processes have shaped the Moon's surface. No agreement on the extent of these processes was possible without the chemical evidence. The existence of titaniferous basalt and, apparently, anorthositic highlands was a surprise to all. Also, though relative ages of many features could be inferred from pictures, only the radioactive dating could give numerical values. The seismic effects are most surprising and lead us to ask what the structure beneath these smooth maria may be. Also, the magnetic effects lead us to ask many questions in regard to lunar history. However, these studies would be even greater if we did not have the exact and detailed pictures that are present in this volume.

We hope that the space program continues and that our satellite does indeed give us an interesting history of the early years of the solar system which appears to be closed to us except for the interesting results to be secured from meteorites and those to be secured from the study of the Moon.

Harold C. Urey

Contents

I. THE MOON—AN ALIEN WORLD

1. Introduction

The recent years of our lives will go down in the history of mankind on this planet replete with milestones marking a rapid sequence of events which still make us gasp in awe, and to which our descendants will look back in wonder. In particular, the date of December 24, 1968 is bound to remain enshrined forever as one of the most memorable landmarks in the age-old quest of human endeavor; for on that day three human beings—the American astronauts, William A. Anders, Frank Borman and James A. Lovell—who, three days before, disengaged themselves from the gravitational field of the Earth, reached the proximity of the Moon in their spacecraft Apollo 8 and allowed themselves to be attached for almost twenty hours to another celestial body. In other words, on Christmas Eve of 1968 the species *Homo sapiens*—a species born and bred on this planet—left for the first time its terrestrial cradle, and took the first steps toward its proliferation throughout the solar system.

And more; for in May 1969—five months later—another three intrepid Americans manning the Apollo 10 spacecraft managed to spend a total of some 140 man-hours in the lunar orbit—a feat that earned its crew, consisting of Eugene A. Cernan, Thomas P. Stafford and John W. Young, the right to claim citizenship of the Moon by naturalization. To be sure, none of the first six men who came from the Earth to take a close-up look at the Moon up to the first half of 1969 actually descended on its surface; but while in orbit around the Moon they could consider themselves lunar, just as the Earth-circling astronauts are terrestrials.

As everyone knows, the first actual descent to the lunar surface was accomplished on July 20, 1969, when two American astronauts of the Apollo 11 mission—Neil A. Armstrong and Edwin E. Aldrin, Jr.—set foot on the vast plains of the lunar Mare Tranquillitatis, while the third—Michael Collins—piloted around the Moon the spacecraft in which all three eventually returned to Earth. Moreover, their feat was repeated only four months later by Alan L. Bean, Charles Conrad, Jr. and Richard F. Gordon, Jr. of the Apollo 12 mission, who landed with equal success on November 19 in the central parts of Oceanus Procellarum. Thus, up to the end of 1969, astronauts of the Apollos 8 through 12 missions spent a total of seven days in lunar captivity, of which sixty-nine man-hours were actually spent on its surface—all in anticipation of greater things to be expected in the future.

Materials traffic in the Earth-Moon system is, to be sure, nothing new in the world; for lunar debris must have fallen on the Earth from time to time in the past. As we shall detail later in the text (Chapter 3), the impact of every major meteorite on the bare surface of the Moon is bound to cause a "kick-off" and spill-over of relatively large amounts of lunar rocks into space. Most of these are eventually bound to be swept up by the broom of terrestrial attraction onto the Earth, where they may reappear in various guises of local naturalization. While the Earth-bound transport from the Moon can thus be kept alive by intermittent natural processes, it is the traffic in the opposite direction which is much more difficult to accomplish, not only because of the six-times-greater terrestrial gravity and thus higher velocity of escape, but mainly because of the presence of our overprotective atmosphere—whose resistance makes escape of solid material beyond the gravitational confines of the Earth virtually impossible. It was not until the advent of man with all his technological ingenuity that this age-old situation at last began to change.

In point of fact, the onset of the age that opened up the exploration of the Moon by spacecraft in the late 1950's came so suddenly as to catch more conservative spirits among us unaware, though the essential technical elements necessary for this purpose were behind the scenes for some time waiting to enter the stage. Three distinct branches of human technology—rocket propulsion, long-range radio communications and computer control, with roots quite independent of each other—had to mature sufficiently for their confluence to make spaceflight an accomplished fact; and we should be thrilled that this came to pass within our lifetime.

Human interest in the Moon and its surface markings, to be sure, antedates by centuries the advent of the space age. The earliest map of the Moon still extant appears to have been prepared by William Gilbert (1540–1603), physician-in-ordinary to Queen Elizabeth I and—his most important claim to fame—discoverer of the terrestrial magnetism. In a book entitled *De Mundo Nostro Sublunari*, which remained unfinished at the time of his death (and did not appear until 1651 in Amsterdam through the good offices of James Boswell), Gilbert drew what appears to be the first map of the Moon (see Figure 1) bearing some resemblance to the lunar face as we know it today. As,

I. THE MOON— AN ALIEN WORLD

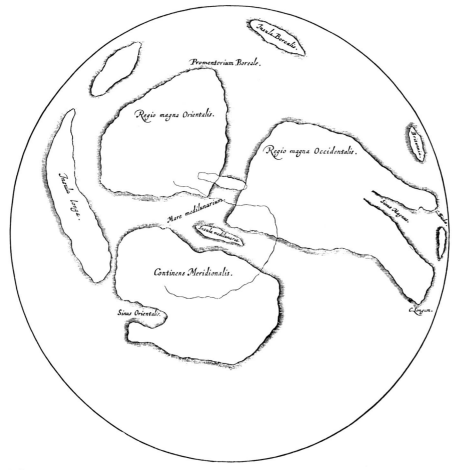

1. A drawing of the face of the Moon based on naked-eye observations by William Gilbert and completed by him some time before 1603. It represents the first as well as the last map of the Moon that has come down to us from the pretelescopic era.

[4]

1. Introduction

moreover, Gilbert died more than six years before the discovery of the telescope, his map of the Moon was the first as well as the last one based on the naked-eye observations of the pretelescopic era.

After the discovery of the telescope in 1609 the Moon became at once one of the principal objects of observation with its aid; and numerous drawings of the lunar face were made by Galileo Galilei (Figure 2). However, a glance at his drawings reproduced herewith should be sufficient to disclose that Galileo was no great astronomical observer, for none of the features recorded on his drawings can be safely identified with any known features of the lunar landscape. Almost simultaneously with Galileo's observations in Italy, the Moon came under telescopic scrutiny in England through the work of Thomas Harriot (1560–1621), a not generally well-known figure among the British astronomers of the Elizabethan era. His drawings (made in 1610) are much superior to Galileo's, as the reader may gather from an inspection of the one reproduced in Figure 3, in which several well-known features of the lunar face are clearly recognizable.

The limitations of early telescopic work—like the limitations of the Russian photographs of the far side of the Moon in 1959, which possessed a resolution comparable with that attained of the front side by the telescopes in Galileo's time—were really unimportant in the long run; for both proved to be forerunners of greater things to come. Unlike what we have witnessed since 1959, in the early days of telescopic astronomy of the Moon progress was slow. And it took more than a hundred years after Galileo's time before the first relatively reliable set of lunar coordinates was established by Johann Tobias Mayer around 1750.

In more recent times—since the second half of the nineteenth century—a gradual increase in our acquaintance with the topography and morphology of the lunar face has been greatly improved and expanded with the introduction of photography in the service of lunar studies; and just before the advent of the space age, ground-based lunar photography with the best telescopic means enabled astronomers to resolve near the center of the apparent disk of the Moon (where direct view is least vitiated by foreshortening) details about 400 meters in size, those smaller still being irretrievably blurred in the haze rising from the diffraction phenomena, unsteadiness of seeing, and photographic plate grain. Two major programs of lunar photography of this type have been carried out in the last decade: one by Gerard Peter Kuiper and his associates at the University of Arizona, the other by the present writer and his associates at the Observatoire du Pic-du-Midi in the French Pyrénées (see Plates 1–3). Both resulted in the publication of two atlases of the visible face of the Moon (Kopal *et al., Photographic Atlas of the Moon,* 1965; Kuiper *et al., Consolidated Atlas of the Moon,* 1968) which constitute probably the last records based entirely on photographs taken from the Earth—before such photography was largely outclassed by data secured from spacecraft employing much smaller optics, but operating in much closer proximity to their target. The decade just past witnessed also a completion of the most extensive mapping project of the visible face of the Moon, undertaken by the Aeronautical Chart and Information Center of the U.S. Air Force on the scale of 1:1,000,000 (largely on the basis of lunar photographs secured from Pic-du-Midi); and the results of this work are likewise available to the general public.

Since 1957, when the first artificial satellites were placed in orbit around the Earth, events started moving with truly dramatic speed; and their results still keep us spellbound. On September 13, 1959—another memorial date in the annals of our science—the first particles of terrestrial matter (in the form of the Russian Luna 2 and the last stage of its carrier rocket) crash-landed on the lunar surface in the southeastern corner of the lunar Mare Imbrium as a harbinger of greater things to come. A month later, Luna 3 disclosed to us the first televised image of the Moon's far side and circumnavigated our satellite —a feat accomplished 437 years (a mere instant in cosmic history) after the

[5]

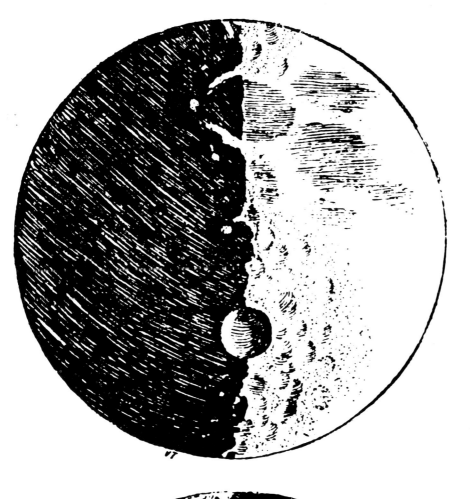

2. *Two of the drawings of the Moon made by Galileo Galilei some time in the second half of 1609 with the aid of a telescope. The drawings apparently reproduce but schematized views of the Moon with Galileo's early telescopes; for none of the features recorded on them can be safely identified with any known formations.*

1. Introduction

first circumnavigation of our own planet by Magellan, which confirmed the spherical shape of the Earth.

Since that time, up to the first half of 1968—which marked the end of an era of unmanned one-way explorations of the Moon—no less than thirty-four additional spacecraft of American as well as Russian origin followed in the wake of those first inanimate messengers to the Moon launched in 1959, seven of which managed to effect soft landings, while ten became for a time artificial lunar satellites (see Tables 1–3 of the Appendix). While the resolution of the images of the lunar far side televised to us by Luna 3 in 1959 was between 20 to 30 kilometers on the surface of the Moon—that is, about the same as attained on the front side by the early telescopes of Galileo—the images televised by the hard-landing Rangers in 1964–65 (see Plate 4) resolved details down to 1 meter in size near their respective points of impact (about 30 centimeters for Ranger 9); and the soft-landers like the American Surveyors (Plates 5 and 6) of 1966–68 have increased this resolution to less than 1 millimeter in size in the immediate neighborhood of the spacecraft.

The greatest contributions to our present knowledge of the morphology of the lunar surface were, however, made by the photographic cameras of Lunar Orbiters in 1966–68 (see Plate 9). This is not the place in which to give a description of the technical aspects of these spacecraft. (For appropriate sources see the Bibliographical Notes at the end of this volume.) Suffice it to say that the entire surface of the Moon—its front as well as its far side, some 38 million square kilometers in area—was photographed by them (largely from overhead vantage points, obviating the effects of foreshortening) with a ground resolution of 100 to 200 meters; while almost 100,000 square kilometers of the lunar front side was photographed by the Orbiters' high-resolution optics with a ground resolution of a few meters. As a result of this highly successful program, we now possess a photographic coverage of the lunar surface almost as complete as we possess for our own Earth; and the Orbiter

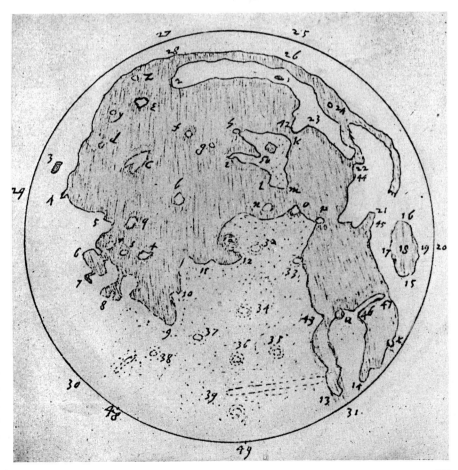

3. The first telescopic map of the full Moon prepared by Thomas Harriot in England some time in 1610. It is much superior to Galileo's drawings, and contains many features that can be clearly recognized today (the reader should note that Harriot as well as Galileo used telescopes which did not invert the image).

[7]

films (televised from the Moon and reconstituted on the Earth), equivalent in extent to more than eight miles of 35-mm film, constitute a vast mine of new data whose interpretation will keep students of the Moon busy for many years.

The last of the photographic Orbiters ended its mission (like all its predecessors; see Table 3) by crash-landing on the lunar surface on January 29, 1968; and the last of the soft-landers (Surveyor 7) gave up the ghost about a month later, thus bringing to a close the glorious period of 1959–68 of unmanned exploration of our satellite—an effort that provided all essentials for the next phase of lunar exploration characterized by the entrance of man on the stage. With the advent of man, the acquisition of lunar photographs from space became considerably simplified. All previous views of the Moon recorded by the Rangers, Surveyors or Orbiters had to be televised down to the Earth and reconstituted in the terrestrial laboratories on the ground (a process in the course of which all Orbiter photographs acquired their characteristic striplike structure). With the entry of man on the scene and his return to the Earth with all the results of his work, we have acquired directly the first-generation negatives of the lunar surface—of the kind which once existed aboard each one of the Lunar Orbiters, but which perished with them as each spacecraft of this class crash-landed on the lunar surface in order not to interfere with the communication systems of its successors.

The first lunar spacecraft which returned to us on Earth with the negatives of its original photographs taken in lunar proximity was the Russian Zond 6—still unmanned—which was launched from Tyuratam, USSR, on November 10, 1968, and recovered on November 17, after a soft landing inside Soviet Russia. However, the principal contributors to the import of first-generation negatives from the Moon so far have been the American missions Apollo 8 and Apollos 10 through 12, and the Hasselblad cameras in the hands of their astronauts, both in orbit around the Moon and (for Apollos 11 and 12) on its surface.

With the advent of this new era in lunar-terrestrial relations, which the current advances in space technology have so suddenly thrust upon us, it may be timely to take stock of our present state of acquaintance with the topography and structure of the lunar surface in the form of an atlas, which should summarize our present knowledge in pictorial form, addressed to a nontechnical reader. To be sure, many maps and atlases of the Moon have appeared in the past, both near and distant; but even the most recent ones are fundamentally deficient in one important aspect, namely, they still present the Moon as seen from the Earth in terms of the terrestrial photography.

And yet the recent contributions of space photography—in particular, of the American Orbiters 1 through 5 between 1966 and 1968—have truly revolutionized the subject, and their importance cannot be overestimated. For not only did they attain a spatial resolution a thousand times as high as that obtainable with ground-based telescopes from the distance of the Earth (a million times as high in the close proximity of the soft-landers); but they also provided photographic coverage of the entire globe of the Moon—of its front as well as far side—from very different vantage points; and thus did largely away with the effects of foreshortening which vitiates the view of the apparent lunar disk, as seen from the Earth, toward the limb.

The aim of the present atlas will be to present the Moon in all its aspects to the reader, through a judicious selection of ground-based as well as space-borne photographs of recent origin. All terrestrial photographs included in this atlas have been secured since 1964 from the Observatoire du Pic-du-Midi at an altitude of 2,862 meters—mainly with the observatory's new 43-inch reflector, as part of an extended collaborative program between the University of Manchester and the U.S. Air Force, carried out under the direction of the author of this book, with the wholehearted support of the observatory's director, Professor Jean Rösch. Most of these photographs—taken by M. T. Jones, T. W. Rackham, but mainly by P. V. Sudbury and B. Temple at the 43-inch reflector—are being published for the first time, while most of the

[8]

1. Introduction

darkroom work and preparation of the prints for this book was carried out at Manchester by Charles Lowe.

All space-borne photographs included in the atlas have been secured as part of the American Ranger, Surveyor, Orbiter, and Apollo projects; and the assistance of NASA, the U.S. Air Force, as well as of the Aerospace Division of the Boeing Company in Seattle (the home of the Lunar Orbiters) in making available the basic data is hereby gratefully acknowledged.

The tremendous amount of data now open to students of the Moon called, of course, for a limited selection of available material to compress the present atlas to a manageable size of a little over 200 plates. Partly for this reason, we gave up any effort—characteristic of most lunar atlases of earlier date—to have all parts of the Moon uniformly covered by photographic evidence. To do so at the spatial resolution made possible by the Orbiter photography would have called for a work consisting of many volumes of the size of the present atlas; and such a work, while desirable, must definitely remain a task for the future.

Moreover, even so severe a selection of the data as could be accommodated in the present volume could scarcely be presented in a form attractive to a wider community of users unless the photographic evidence were accompanied by an adequate amount of supporting text, aiming to explain its meaning within the framework of our present knowledge of the physics and astronomy of the Moon. Without the accompanying text, we feel that a mere perusal of a collection of 200 or more photographs could hardly retain for long the attention of the reader who is not an expert in lunar studies and does not already know his way about the Moon.

On the other hand, a combination of documentary photographs with explanatory text should be able to arrange for the reader a pictorial guided tour of the different landmarks of interest on the lunar surface, and explain to him by words as well as pictorial evidence the meaning and significance of the principal types of formations encountered on the Moon, which are so different from almost anything that we know on the Earth. In point of fact, the present work should be regarded as a kind of lunar Baedeker, which can accompany its user in his own explorations of the Moon from the safety of his armchair, and provide him with an added sense of participation in the different episodes of lunar exploration which he can follow from a safe distance on the screen of his television set.

Not many men will reach the Moon in the flesh and return home safely within our lifetime. Yet we, the terrestrial plodders—whom age or other considerations may have relegated to the role of onlookers at the heroic deeds now unfolding before our eyes—should rejoice at the thought that these modern voyages of discovery outside the gravitational boundaries of our Earth should have come to be made within our lifetime. That this happened is an accident of history for which we should be duly grateful.

2. The Moon—Our Nearest Celestial Neighbor

The aim of the present chapter will be to introduce to the reader the Moon as a celestial body, and to review briefly its basic properties—its distance from us and characteristics of its motion, and physical properties like its size, mass and density—a knowledge of which should provide the necessary background for understanding much of what we see on its surface. To the astronomers, to be sure, the Moon has been an old friend from time immemorial; and at least a rudimentary knowledge of its motion goes very far back in the history of mankind, for since prehistoric times the waxing and waning of lunar phases and the light changes accompanying them provided the first astronomical basis for the reckoning of the time. Whenever we delve sufficiently far back in the history of almost all primitive civilizations, we find them invariably dependent on the lunar, rather than on the solar, calendar: the month became a unit of time long before the concept of the year emerged from accumulating observations; and the Moon as the graceful carrier of this knowledge thus gained entrance, as a female deity, to the pantheons of most ancient nations.

Perhaps the most important single characteristic of the lunar relative orbit around the Earth is its size, that is, the distance separating us from our only natural satellite. How far is it to the Moon? A determination of its mean distance does not represent any recent acquisition to our knowledge, for at least its order of magnitude was known to Aristarchos of Samos some three centuries before Christ; and Hipparchos of Nicaea determined it to within a few per cent of its correct value from the relative duration of the successive phases of lunar eclipses.

In more recent times, astronomers have redetermined this distance more accurately by a method to which we resort whenever we wish to measure the separation of any inaccessible point on land or sea, namely, triangulation. The astronomical triangulation of lunar distances using the known terrestrial diameter as a baseline revealed that the relative orbit of the Moon around the Earth is (approximately) an ellipse, with a radius-vector varying between 364,400 and 406,730 kilometers. The mean distance of the Moon in the course of a month amounts, very closely, to 384,400 kilometers and is equal to 62.267 times the Earth's equatorial radius (which would, therefore, be seen from the center of the Moon at an angle of 57'2".7, called the Moon's mean horizontal parallax), or 0.00257 times the mean distance separating us from the Sun (the latter being, on the average, 389 times as far from us as the Moon). The distance to the Moon represents, therefore, less than 1 per cent of the distance separating us from our next two nearest celestial neighbors—the planets Venus and Mars—even at the time of their closest approach. Many a terrestrial traveler has probably accumulated, during his lifetime, a greater mileage by automobile than would be involved in a round-trip to the Moon. Light traverses this distance in 1.28 seconds; and an average spacecraft which disengages itself from the gravitational field of the Earth can reach the Moon after a free flight of sixty-five to seventy hours.

Quite recently, physicists sending out radar pulses to the Moon and timing accurately the return of their echoes reflected from the lunar surface redetermined the distance to the Moon by a laboratory method relying on a knowledge of the velocity of propagation of radio waves through empty space (nowadays known to better than one part in a million). This method, strictly speaking, measures the instantaneous distance between the transmitting-receiving antenna on the terrestrial surface and the reflecting patch on the Moon. The time variation in this range, however, permits us to separate the size of the two bodies from the distance of their centers. In 1958, Yaplee and his collaborators redetermined in this way the distance to the Moon and found its mean value to be $384,402 \pm 1$ km, in close accord with previous astronomical determinations. The distance separating us from the Moon at any time can therefore nowadays be regarded as known to within approximately 1 km; and further refinements of the radio-echo techniques (coupled with better knowledge of the exact shape of the Moon) are potentially capable of reducing the remaining error by a factor of five to ten.

Still more recently, radar echoes are being superseded for this purpose by laser echoes—or brief flashes of coherent light which are visible to the naked eye. The first laser contact with the Moon was established experimentally in 1962 from the Lincoln Laboratories of the Massachusetts Institute of Technology by illuminating a small spot on the Moon with a pencil of coherent light, a tiny bit of which was diffusely back-scattered to the Earth; and there was the laser contact established by Surveyor 7 in January 1968. The cube-corner reflector installed on the Moon by the Apollo 11 astronauts in July 1969 is now making it possible to obtain specular reflection of terrestrial laser signals from a much smaller area of the lunar surface—a facility that promises to open up an entirely new chapter in our knowledge of the motion of the Moon in space, as well as its motion around its center of gravity.

The orbit of the Moon around the Earth is approximately an ellipse —of eccentricity $e = 0.05490$—and its plane is inclined to the ecliptic (that is, the plane in which the Earth revolves around the Sun) by an angle i whose mean value as of 1969 was equal to $5°8'43''.4$. The actual lunar orbit departs, however, from an exact ellipse, because its form is distorted by the attraction of the Sun (or, more precisely, the difference of the solar attraction on the Moon and the Earth); and also (though to a much smaller extent) by the oblateness of the Earth and the attraction of other planets.

The principal effect of solar perturbations is to cause the Keplerian ellipse of the lunar orbit to rotate slowly in space in such a way that the line of apsides, joining the points at which the Moon is nearest to the Earth (perigee) and farthest away from it (apogee), advances (that is, moves toward the east) at a rate of $146'427''.9$ per annum (the rest of the perturbations add only $8''.3$), corresponding to a complete revolution of the apsidal line in 3,232.57 days or 8.85 years. In the meantime, the line of the nodes (the line in which the lunar orbit intersects the ecliptic) is receding (moving west) at an annual rate of $69'679''.4$—that is, a little less than half the rate at which the apsides are advancing—corresponding to a complete revolution in approximately 18.6 years.

This motion of the lunar orbit in space has several important consequences, and the most immediate ones concern the orbital period of our satellite. If we define this period as an interval of time after which the Moon will have made a complete revolution with respect to the celestial sphere— the so-called sidereal month—this is known to be equal to 27.321661 mean solar days, or approximately $27^d7^h43^m11.5^{sec}$. During the time of one sidereal month, however, the Sun has moved eastward by approximately one-twelfth of the

2. The Moon—Our Nearest Celestial Neighbor

entire circle; and the Moon would consequently have not returned to the same phase relative to the Sun. As a result, the time interval which will elapse between the two successive identical phases of the Moon—the so-called synodic month—is longer than the sidereal one, and equal to $29^d12^h44^m2.8^{sec}$; this is the time interval in which the Moon always returns to the same phase. However, should we define a month as a time interval in which the Moon will return to the same place in its relative orbit around the Earth, this month—the so-called anomalistic month—will likewise be longer than the sidereal one (though shorter than the synodic month) because of the secular advance of the apsides; and for the known rate of this advance it is equal to $27^d13^h18^m37.4^{sec}$; while a time interval between two successive nodal passages—the so-called draconic month—is shortest of them all (because the nodes regress) and equal to $27^d5^h5^m35.8^{sec}$.

I hope the reader will bear with this abundance of technical terms characterizing the various periods of lunar motion. Their determination represents a remarkable achievement of science; and the length of the sidereal month (the most important of them all), based as it is on astronomical observations extending over many centuries, constitutes one of the most precise measured quantities known to our science. We can state it now to not less than twelve decimal places. Moreover, certain relations which accidentally exist between the lengths of the synodic and draconic months are of interest for the predictions of the solar or lunar eclipses. If the draconic and synodic months were identical, then each new or full Moon would occur in the same relative position with respect to the nodes; and either we should have at each new Moon a solar eclipse (and at each full Moon a lunar one), or none at all. Observations disclose that this is not the case, for eclipses occur more rarely (a few per year at most), a fact which alone proves the motion of the nodes. However, it happens that 223 synodic months are almost equal to 242 draconic months (the difference between the two multiples being only $51^m41.2^{sec}$); so that the positions of the Moon relative to the nodes should be almost the same every 6,585 days, or eighteen years and ten to eleven days (depending on whether this interval includes four or five leap years). In consequence, should an eclipse of the Sun or the Moon occur at a given time, it should recur at the same place after a time interval of 6,585 days—a period already known to the Chaldeans twenty-four centuries ago, and denoted by the Greeks as Saros. Even closer is the coincidence of 716 synodic and 777 draconic months, leaving a difference of only $9^m46.1^{sec}$. Therefore, the eclipses should recur more closely after an interval of 21,144 days or just under fifty-eight years; and still longer periods exist after which this is even more accurately the case.

After having mentioned the principal secular perturbations of the relative lunar orbit—the advance of the apsides and regression of the nodes—caused by solar attraction, a few words should be said on periodic perturbations due to the same cause. The most important periodic perturbation of the Moon's motion is the so-called evection, due to the effect of solar attraction on the apparent eccentricity of the lunar orbit. If the apsidal line is parallel with the direction toward the Sun (that is, when the Moon is full at perigee and new at apogee, or vice versa) the attraction of the Sun tends to elongate the ellipse of the relative lunar motion and increase its eccentricity. If, on the other hand, the apsidal line is perpendicular to the line joining the positions of full and new Moons—a line which astronomers refer to as a syzygy—solar attraction will tend again to widen the ellipse and diminish its eccentricity. In all, this phenomenon, recurring in the period of 31,807 days, can displace the position of the Moon in longitude by as much as $1°16'26''.4$, and cause the apparent orbital eccentricity (the mean value of which is, as has already been mentioned, equal to 0.0549) to fluctuate between 0.044 and 0.066. The displacement in longitude due to evection, which can amount to five apparent semidiameters of the Moon, was discovered in antiquity and known to Hipparchos.

The second most important perturbation—the so-called variation—arises from the periodic interference of the attraction of the Sun and the Earth. Because of their respective relative positions, the gravitational pull of these bodies acts in the same direction if the Moon is approaching the full or new phase, but against if the Moon is perpendicular. The period of this variation will therefore be one-half of the mean synodic month; but although its action can displace the position of the Moon in longitude by as much as $39'39''.9$ (more than its apparent semidiameter) it remained unnoticed to all Greek astronomers including Ptolemy, perhaps because it vanishes (that is, changes sign) at both full and new Moon, and does not affect the times of solar or lunar eclipses with which the Greeks were so much concerned. It was apparently first noticed by the medieval Arab astronomer Abdul Wefa (940–998) and established by Tycho Brahe (1546–1601). Another periodic perturbation of the Moon's motion—third in order of importance because of its magnitude —is the so-called annual inequality, due to the eccentricity of the terrestrial orbit around the Sun. Its amplitude attains $11'8''.9$ in longitude, and fluctuates in the period of one (anomalistic) year. It was discovered by Johannes Kepler (1571–1630).

All the phenomena listed above give rise to perturbations in lunar motions which are sufficiently large to be noticed by the naked eye; and all of them were discovered in the days of pretelescopic astronomy. Their understanding was, of course, impossible until Isaac Newton laid down the principles of the celestial mechanics toward the end of the seventeenth century. The first analytical theory of the Moon's motion was attempted by Newton himself. He succeeded in accounting for the existence of the principal lunar secular as well as periodic perturbations mentioned in this section. Even so great a brain, however, found the problems posed by the theory of the Moon's motion more than a match for his ingenuity; and he is said to have told his friend Edmund Halley that it "made his head ache and kept him awake so often that he would think of it no more."

Through cumulative efforts of many distinguished mathematicians and astronomers in the last three hundred years—from Leonhard Euler (1707–1783), Joseph Lagrange (1736–1813), Pierre Simon de Laplace (1749–1827) to Henri Poincaré (1854–1912), to name only the greatest—the theory of the motion of the Moon (predicting its apparent position in the sky at any time) has gradually been developed into an instrument so precise that the Moon's observed motion has now universally been adopted as the fundamental astronomical clock marking the passage of time. It is true that crystal or molecular clocks in the laboratory can now measure short intervals of time with greater precision than that attainable by any kind of astronomical measurements; but an inherent lack of stability of such devices over long intervals renders our Moon still the most reliable guardian of time. In this respect, the Moon has retained its traditional role from the dawn of civilization to the present, and is likely to do so for a long time to come.

The importance of our only natural satellite in this connection can thus scarcely be overestimated; and it should be realized that the Moon has retained this importance only because its motion has been so thoroughly understood—thus making possible a sufficiently accurate conversion of the observed positions into time. In point of fact, our astronomical time is defined now as the independent variable in the lunar equations of motion; and the observed motion of the Moon thus serves as a hand of the universal clock.

We should, perhaps, also add that all complexities of lunar motion which proved a headache for Newton (and for others also) are not attributable to the Moon itself, but rather to the external influences (mainly the attraction of the Earth and the Sun) to which our satellite is so completely exposed. On account of the smallness and nearly spherical distribution of its mass, the Moon has virtually no control of its motion through space. It is almost completely at the mercy of outside forces which it cannot influence.

[14]

2. The Moon—Our Nearest Celestial Neighbor

With the shape and dimensions of the lunar orbit thus known, we can easily evaluate the mean velocity of its relative motion around the Earth. It averages out to 3,681 km/hour, or 1.023 km/sec—corresponding to a mean angular velocity on the celestial sphere of about 33′/hour (which is just a little more than the apparent diameter of the Moon itself).

The revolution of the Moon around the Earth, and then with the Earth around the Sun, are not the only motions performed by our satellite. It also rotates about an axis fixed in space and inclined by approximately 83°20′ to its orbital plane, with a uniform angular velocity in exactly the same period as it revolves around the Earth, thus showing us almost exactly the same face each month. The period of axial rotation coincides now with the sidereal month within 0.1 second (that is, approximately one part in thirty million) and may have been so synchronized (by the action of tidal friction in the Earth-Moon system) for an astronomically long time. Moreover, the lunar axis of rotation, the pole of the ecliptic and that of the lunar orbit are all situated in the same plane and in that order (see Figure 4) as was established by G. D. Cassini in the form of empirical laws bearing his name. The angle i between the Moon's orbit and the ecliptic is known with great exactitude to be equal to $5°8′43″.4$. The inclination I of the lunar equator to the ecliptic is known with somewhat lesser precision to amount to $1°32′±1′$ (mean of many determinations). Hence, in accordance with Cassini's third law, the inclination of the Moon's equator to its orbital plane should be equal to the algebraic sum of $I + i = 6°40′$.

The face of the Moon seen from the Earth appears to be always the same because of the synchronism prevailing between rotation and revolution, but is not exactly so, for several reasons. The first is the fact that whereas the axial rotation of the Moon is uniform, the angular velocity of revolution in an elliptical orbit (varying as it does with inverse square of the radius-vector) is sometimes ahead, and sometimes behind the orbital motion—causing angular displacement of lunar objects in longitude by as much as 7°54′ about the center of the Moon, a phenomenon known as the optical libration in longitude (see Figure 5).

Secondly, as the lunar axis of rotation is not perpendicular to the orbital plane, but deviates from 90° by $I + i$, sometimes we can see more of one polar region and at other times more of the other in the course of each month. This gives rise to an optical libration in latitude, by approximately 6°50′. Again, when the Moon is rising for the observer on the Earth, we look over its upper edge, seeing a little more of that part of the Moon than if we were observing it from the center of the Earth; and when the Moon is setting the converse is so. This diurnal libration (not of the Moon, strictly speaking, but rather of the observer) amounts to approximately 57′.1 (just under 1°), and superposes upon all other librations—optical as well as physical (small oscillations of the Moon as a rigid body about its center of gravity, invoked by the couple of attracting forces due to the Earth and the Sun, whose selenocentric amplitudes are generally less than 1′)—to enable us to see considerably more than one-half of the lunar surface from the Earth. On the whole, not less than 59 per cent of the entire lunar globe can be seen from the surface of the Earth;

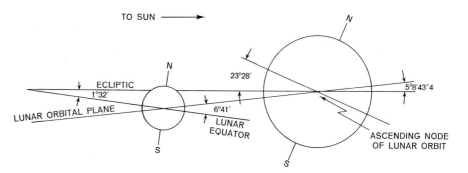

4. Relative orientation of the equatorial planes of the Earth and the Moon with respect to the ecliptic.

only 41 per cent being permanently invisible, and 41 per cent never disappears; the remaining 18 per cent is alternately visible and invisible.

The apparent diameter of the lunar disk in the sky has since ancient times been known to be close to ½° (Archimedes and Hipparchos adopted for it the round value of "720th part of the zodiac"), and varies somewhat because of the varying distance of the Moon from the observer on the surface of the Earth as well as on the position in its relative orbit. When the topocentric observations are freed from the parallactic effect and reduced to the center of the Earth, the mean geocentric apparent diameter of the lunar disk is found to amount to 1865″2—oscillating by 204″8 between perigee and apogee (see Figure 6)—which at the mean distance of 384,402 km corresponds to a mean radius of the lunar globe of 1,738 km. The Moon is, therefore, little more than one-quarter the linear size of the Earth. Its surface covers an area of 37.96 million km²; and its volume amounts to 21.99 million km³ (approximately 2.03 per cent of the volume of the Earth).

The next quantity of basic importance for the understanding of the fundamental characteristics of the Moon is its mass. The mass of the Moon—like that of any other celestial body—can be determined only from the effects of its attraction on another body of known mass; and in the case of the Moon, this will be our Earth (or, quite recently, an impinging spacecraft). There are, in principle, three ways by which this can be done, that is, by a study of the effects of the Moon on (a) the orbital motion of the Earth in space; (b) the axial rotation of the Earth; (c) the infall of a spacecraft.

To take up (a) first, the oft-repeated statement that the Earth revolves around the Sun in an ellipse is not sufficiently precise, for, in actual fact, it is the center of gravity of the Earth-Moon system which describes (approximately) this curve; while both the Earth and the Moon revolve around this common center of gravity in orbits which are exactly alike in form, but whose absolute dimensions are inversely proportional to their masses.

The Earth does not, therefore, revolve around the Sun in an ellipse; it rather wobbles about it in the period of one (synodic) month; and so does, of course, the Moon—only more so. In point of fact, inasmuch as the rate of free fall toward the Sun at the mean distance of the Earth is twice as large as that of the fall of the Moon toward the Earth, the curvature of the absolute orbit of the center of gravity of the Earth-Moon system around the Sun is twice as large as that of the lunar orbit around the Earth. It follows, therefore, that the absolute orbit of the Moon in space is always concave relative to the Sun—at the time of the new Moon it is merely smaller than at full Moon—but its curvature never changes sign. This particular characteristic of its space motion, incidentally, renders our Moon unique among all satellites of the solar system; as it is also in several other dynamical respects, which lack of space prevents us from discussing in detail.

5. A comparison of different aspects of the face of the full Moon, influenced by optical librations of our satellite.

2. The Moon—Our Nearest Celestial Neighbor

But let us return to the monthly wobbling of the Earth in space because of the presence of the Moon. The necessary result of such a wobbling must be a slight alternate eastward and westward apparent displacement, on the celestial sphere, of every astronomical object viewed from the Earth, as compared with the place where this object would be seen if the Earth had no satellite and moved around the Sun alone. In the case of the stars (or of more distant planets) this displacement is insensible; but it can be easily identified in the apparent motion of the Sun—or, better still, of one of the nearer planets or asteroids. From extensive observations of this kind it has been established that the mean distance of the Earth's center from the common center of gravity of the Earth-Moon system amounts to 4,667 km. This places the center of gravity of the Earth-Moon system well inside the terrestrial globe (though nearer to its surface than to the center), and represents about one part in 82.37 of the mean distance separating the centers of the Earth and the Moon. By elementary principles of mechanics it follows that the mass of the Moon is equal to $1/(82.37 - 1) = 1/81.37$ times that of the Earth. According to the latest analysis by Dirk Brouwer and G. M. Clemence (1961), the mass-ratio Earth:Moon was found by this purely astronomical method to be equal to 81.37 ± 0.03—subject to an uncertainty of about one part in three thousand.

Quite recently, this and other astronomical determinations of the Earth: Moon mass-ratio were largely superseded by more direct determinations of the Moon's mass from the accelerations imparted by it on man-made fly-by or hard-landing spacecraft (see Table 1). Thanks to the accurate range-doppler tracking of such spacecraft (in particular, of the hard-landing Rangers), we know the Earth:Moon mass-ratio to be equal to 81.303 ± 0.001. This latter value renders the (mean) distance of the Earth's center from the common center of gravity of the Earth-Moon system equal to 4,671 km. Moreover, the absolute masses of the Earth and of the Moon result as $m_\oplus = 5.978 \times 10^{27}$ g and $m_\mathbb{C} = 7.353 \times 10^{25}$ g, respectively; their uncer-

6. *Difference in the apparent diameter of the Moon between perigee and apogee, due to the eccentricity of the lunar orbit.*

[17]

tainties of the order of one part in a thousand are almost solely due to the limits of our present knowledge of the gravitation constant.

The lunar mass of 7.35×10^{25} g, or over 73 trillion metric tons, may loom large in comparison with all terrestrial standards; but on cosmic scales it constitutes but a relatively tiny speck. And neither is the mean density of the lunar globe at all unusual; for dividing the mass just found by the lunar volume of 2.199×10^{25} cm³ we find its mean density to be $\varrho = 3.34$ g/cm³—that is, only a little more than the density of common granite rocks of the Earth's crust (2.78 g/cm³), and considerably less than the mean density of the terrestrial globe (5.52 g/cm³). The gravitational acceleration on the lunar surface is, therefore, only some 162 cm/sec² (less than one-sixth of the terrestrial one); and the velocity of escape from the lunar gravitational field is close to 2.38 km/sec—in comparison with its terrestrial value of 11.19 km/sec.

The much reduced gravity prevailing on the lunar surface is the reason the Apollo astronauts were able to swing their heavy backpacks with the life-support systems with relative ease, and why the periods of the lunar-orbiting satellites (see Table 3) are so much longer than for similar circumterrestrial spacecraft. However, low gravity also entails some disadvantages. Although much less muscular work is indeed required on the Moon to lift weights or throw stones, our own weight would work less effectively for us if we wished to use it to compress anything, or to drive a shovel into the ground by stepping upon it.

The relative smallness of the mass of the Moon and the low velocity of escape from its gravitational field entail several important consequences; and perhaps the most important one for an understanding of lunar surface features is the well-nigh complete absence of any atmosphere, which would protect this surface from a direct contact with outer space. Why should a self-gravitating astronomical body of lunar or planetary size possess an atmosphere? While, for the terrestrial planets of masses comparable with that of the Earth, their present atmospheres may constitute mixtures of primordial gases with those liberated from their interiors by essentially thermal processes in the course of their long cosmic past, it is most unlikely that a body as small as the Moon could have permanently retained any primordial gas, so that any atmosphere which it could possess would be regenerated, or possibly accreted.

In order to appreciate why this should be so, let us recall that the continued existence of an atmosphere around any celestial body and its composition testify to the extent of a stalemate between two opposing tendencies: the attraction of the central body weighs on each gas molecule in the same way as on any macroscopic object and prevents the escape of all those whose velocity is (for the Moon) less than 2.38 km/sec, while, on the other hand, the heat pumped in our gas by the Sun (as well as by the surface of the respective planet) maintains the kinetic energy of the gas particles and thus keeps the atmosphere distended.

In general, the velocity of the molecules constantly bouncing against one another will be greater the higher the temperature of the gas; though heavier molecules will require higher temperatures to reach the escape velocity than the light ones. This velocity of escape depends, in turn, on the mass as well as on the radius of the astronomical body in question (being, in general, greater for heavy or dense bodies than for light or distended ones). To give some numerical examples, the escape velocity from the gravitational field of Jupiter—the greatest and most massive planet of our solar system—is more than 61 km/sec. For the Earth this velocity turns out to be 11.2 km/sec, and for the Moon only about 2.3 km/sec. On the other hand, thermal agitation corresponding to a mean temperature of Jupiter's daylight hemisphere of about 130°K (that is, −143°C) is barely sufficient to endow even hydrogen—let alone heavier molecules—with a mean velocity of 1.3 km/sec, which is quite insufficient for any kind of escape even over astronomically long intervals of time.

[18]

2. The Moon—Our Nearest Celestial Neighbor

For the Earth and the Moon, which enjoy warmer climates (on account of their greater proximity to the Sun), the mean velocities of thermal agitation of different molecules would be proportionally higher. Thus, at 0°C, the mean velocities of the hydrogen molecules or helium atoms are found to be 1.84 and 1.31 km/sec, respectively; for molecular nitrogen or oxygen the respective values become 0.49 and 0.46 km/sec, while for water vapor or carbon dioxide the corresponding figures are 0.62 and 0.39 km/sec, respectively. Moreover, if the gas temperature should rise to 100°C (actually encountered on the Moon at the time of high noon), all the foregoing velocities would be increased about 1.17 times.

These all are, to be sure, mean velocities, possessed by most, but not necessarily all, molecules of gas in a potential atmosphere. Some of them can, at any time, temporarily acquire by particularly energetic (or weak) collisions chance velocities considerably higher (or lower) than the average—possibly higher than the velocity of escape from any planet. The number of such particularly fast particles bears, however, a constant ratio to the total number of all gas molecules, predictable statistically for each temperature from the kinetic theory of gases. Hence, it is possible to estimate theoretically a fraction of the molecules or atoms of any gas that are in a position to escape from the atmosphere of any planet in a given time. A more detailed analysis reveals that if the mean molecular velocity of thermal agitation is approximately one-third that of escape, one-half of the corresponding atmosphere will be dispersed in a few weeks; should it amount to one-fourth, the half-life of our atmosphere would be increased to about 50,000 years; while if this ratio were one-fifth, the atmosphere would last for 100 million years.

On the basis of these results we are led to expect that the large and massive outer planets of our solar system—Jupiter, Saturn, Uranus and Neptune—are sufficiently cool to retain even the lightest gases (such as H_2 or He) practically for all time; their high escape velocities (61, 37, 22 and 25 km/sec, respectively) render them virtually foolproof against any leak into space. This inference has indeed been confirmed by spectroscopic observations, which reveal the presence of vast atmospheres around these planets, containing large amounts of hydrogen and its compounds.

As for the Earth, its gravitational attraction proves too weak (and its mean temperature too high) to retain permanently the lightest gases in its atmosphere. Any free hydrogen or helium will "leak" out into space at a (cosmically) quite rapid rate. On the other hand, heavier gases—such as molecular nitrogen or oxygen—can be retained almost indefinitely. The chemical composition of our atmosphere may have changed and evolved profoundly in the course of the long astronomical past of the Earth; but there appears never to have been any dearth of volatile elements over its surface.

When, however, we return to the Moon, whose mean temperature should not differ greatly from that of the Earth (as both these bodies are in the mean equally distant from the Sun and receive from it, on the whole, equal amounts of heat) but whose escape velocity is only about 2.4 km/sec (almost five times less than on the Earth), a different situation is encountered. On the daylight side of the Moon, where (as we shall discuss in more detail later) temperatures as high as 400°K are temporarily attained, the lightest gases—hydrogen and helium—would disappear into space almost immediately; at least more exact computations reveal that a hypothetical atmosphere of hydrogen (atomic or molecular) around the Moon would keep dissipating at a rate at which its density would diminish to approximately 37 per cent (or, more exactly, to a reciprocal value of the base of natural logarithms) about every ten minutes; and in less than a day at lunar nighttime when the temperature drops below 100°K.

An atmosphere consisting of helium would similarly dissipate in some three hours of daylight, or 200 hours at night. Atomic oxygen or water vapor would take years for dispersal at daytime, and a time comparable with the age of the Moon (10^9 years) at continuing nighttime conditions. Molecular oxygen could last in lunar daylight for millions of years, and carbon dioxide still longer. However, some of those just mentioned (oxygen, for instance) are so reactive that in any event they would not stay long in free state if in contact with a solid surface, but would form compounds in the surface layer of solid rocks. In all, the rate of dissipation into space (or the formation of solid compounds on the surface) of all but the heaviest (or inert) gases—which are again cosmically very scarce—is so high on the Moon that we should not expect to find any appreciable permanent atmosphere around it; and this expectation has indeed been borne out by all aspects of the observational evidence available to us so far.

Perhaps the most direct observation which pointed to the absence of any air around the Moon since the earliest days of telescopic astronomy has been the perpetual absence of any visible clouds over its face. Moreover, no evidence of refraction of light has ever been found at the Moon's limb whenever our satellite places itself between us and a more distant object. During an eclipse of the Sun, the outline of the solar limb remains completely free from any distortion where the Moon intercepts it. Furthermore, whenever on its apparent journey through the sky the Moon happens to occult a star, the latter's light is seen to vanish instantaneously (or as suddenly as is permissible by the laws of diffraction and the angular diameter of the star), and does not fade away gradually as it would if it were dimmed, selectively, by extinction in any perceptible lunar atmosphere.

This absence of refraction is by itself a sufficiently stringent test to enable us to conclude that the density of a hypothetical lunar atmosphere—if any—above the surface must be less than one part in ten thousand of the terrestrial air density at sea level. In more recent years, this upper limit has further been lowered by repeated but so far fruitless quests for an indication of twilight phenomena which should be produced in a hypothetical lunar atmosphere during sunrise or sunset. A quest for the detection of such phenomena—in particular, the polarization of sunlight scattered on the molecules of a hypothetical lunar atmosphere toward the Earth above the cusps of a lunar crescent—has led so far to completely negative results, and depressed the possible upper limit for the density of a hypothetical lunar atmosphere to 6×10^{-10} of the terrestrial one. If the Moon possesses a gaseous envelope, its density on the surface cannot, therefore, exceed some 7×10^{-13} g/cm³ if it is not to produce detectable twilight phenomena; and the actual amount of gas below this limit still remains largely conjectural.

A gas density of the order of 10^{-12} g/cm³ would represent a pretty hard vacuum from the point of view of the terrestrial physicist, and one which is, incidentally, attained in our own atmosphere at an altitude of approximately 180 km above sea level. However, even at such great heights, the number of gas particles remains still of the order of 10^{10} g/cm³; and although the mean free path of such particles between mutual collisions is of the order of 100 m, even so rarefied a gas could manifest itself in different observable ways. It would not, to be sure, offer any protection to the surface beneath from impinging meteorites. A hypothetical lunar atmosphere of surface density of the order of 10^{-12} g/cm³ would not decelerate enough any meteoritic material—small or large—to cushion its impact significantly. All solid particles in space intercepted by the Moon must hit its surface essentially with their original cosmic velocities, and spend themselves by impact on the surface rather than in their passage through the atmosphere as is the case on the Earth.

However, if we return to our own atmosphere such as exists at an altitude of 180 km above sea level, even though it is insufficient to affect the velocity of the meteors passing through it, it can give rise to other interesting phenomena. Between 180 to 200 km above sea level we would find ourselves in the midst of the auroral zone, where luminescent gas stimulated by the impact of corpuscular sunrays produces the beautiful displays of "northern

[20]

2. The Moon—Our Nearest Celestial Neighbor

lights." Are there similar aurorae on the Moon? Herzberg pointed out sometime ago that a search for emission spectra of such displays around the bright limb of the Moon might constitute one of the most sensitive tests of the presence of a hypothetical atmosphere of the Moon. No trace of any such emission, however, has so far been detected above the surface of the Moon, though a quest for fluorescent radiation from the Moon led quite recently to some very interesting discoveries on its surface.

So far as our present knowledge goes, however, the Moon does not possess any atmosphere of density in excess of 10^{-12} g/cm³ at its surface; and how far below this limit its actual density happens to be we can only guess. Let us add that for the astronomers concerned with the relative abundances of the elements of cosmic matter, even the upper limit just stated is already becoming rather uncomfortably low, for, after all, there really should be more gas around the Moon than this limit seems to admit.

Why is it not there? J. R. Herring and A. L. Licht pointed out, some years ago, that atoms of heavy gases can be mechanically "blown off" the Moon into space by collision with corpuscular radiation (mainly protons) emitted continuously—and, sometimes, in angry energetic "puffs"—by our Sun. The existence and intensity of this "solar wind" and its occasional gusts usually associated with flares and other sudden disturbances of the solar surface, are now well known from the soundings of deep-space probes, and indirectly attested to by such terrestrial phenomena as polar aurorae and magnetic storms. Herring and Licht argued that the knocking-off power of this solar wind is, in fact, sufficient to remove most of the heavy inert gases from the lunar surface, and thus despoil it even of such scanty vestiges of a gas envelope which its own feeble gravitational attraction would enable it to retain.

More recently, however, E. J. Öpik and F. S. Singer disclosed another and more effective way of gas removal from the lunar surface: by ionization. As long as gas remains neutral, its atoms or molecules can be removed only by collisions—be it with other molecules, or particles of the solar wind. Should, however, the gas particles become ionized and thus acquire a positive electric charge (and Öpik with Singer has shown how easily this can happen to lunar argon and heavier inert gases), the positive charge of the sunlit hemisphere of the Moon acquired by photoionization of the light elements in its crust can remove ions by repulsion far more effectively than could be accomplished by the collisions with particles of the solar wind.

Such gases that are left by these processes to cling to the lunar surface for a limited time do not, therefore, constitute any real atmosphere—in which individual gas particles are balanced by mutual collisions—but rather a transient exosphere in which the individual atoms or ions describe essentially free-flight trajectories in the prevailing gravitational or electrostatic field. Each planetary atmosphere is bound to peter out into such an exosphere on its outer fringe bordering on interplanetary space; but on the Moon its exosphere apparently reaches down to the solid surface itself. As to its probable chemical composition—due partly to captures from interplanetary space and partly by degassing of the lunar surface layers—Öpik estimated recently that it contains approximately 5×10^5 particles/cm³, among which are about 1.2×10^4 molecules of hydrogen, 1.4×10^5 molecules of water, a comparable amount of carbon dioxide and not more than 1×10^4 atoms of inert gases.

If, moreover, the Moon possesses no detectable atmosphere, it cannot, of course, maintain any liquid on its surface. Near the poles, to be sure, depressions may exist which are never reached by direct sunlight (and which are illuminated, at best, by sunlight scattered from adjacent slopes). In such regions, condensed volatile substances may possibly be present in the form of some kind of permafrost; but should they ever evaporate, they are apt to be lost in a very short time. Hence, no liquid—or even solid—water can be present at any spot on the Moon which can be reached by sunlight. The surface of the Moon must, therefore, be regarded as bone-dry, with none of its visible features formed, or even modified, by running water, or freezing and melting water. With both the hydrosphere as well as the atmosphere effectively absent around the Moon as disturbing agents, the fossil record of the lunar surface should possess a *vastly greater degree of permanence* than anything known to us on Earth; and in the next chapter we shall disclose the facts furnished by the Apollo 11 mission confirming that this is indeed the case.

Before we do so, however, it is our intention to exhaust every possibility of atmospheric phenomena on the Moon. If the Moon cannot possess any permanent atmosphere to speak of, could it, perchance, have acquired at times a transient atmosphere, which could have shielded its surface at least temporarily; and thus enabled, for instance, the existence of fluid flow for at least limited periods of the long lunar past? The answer to this question cannot so far be an unqualified no; for, regardless of any possible past period of temporary volcanism or degassing, a temporary atmosphere could also have been "imported" from outside whenever the Moon suffered a collision with a comet.

The mechanical aspects of such collisions will be taken up in the next chapter, in connection with the problem of the origin of the lunar craters. What interests us at present is not the type of scars which such events may have left on the surface, but rather the subsequent fate of the gas, brought in frozen state from cooler parts of the solar system in the nuclei of cometary heads. As is well known, the nucleus of a comet—the only part of its anatomy possessing any kind of permanence—constitutes an iceberg of frozen hydrocarbons and other compounds of moderate molecular weights, which remain in solid state as long as the comet floats freely in space sufficiently far from the Sun, and whose gradual evaporation in more moderate zones of interplanetary climate gives rise to the beautiful, though ephemeral, phenomena of cometary heads and tails. When, however, a cometary nucleus strikes a solid obstacle—such as the surface of the Moon—a conversion of ice to gas should be virtually instantaneous. Cometary impacts on the Moon must have occurred many times during its long astronomical past and the question arises as to the fate of the gas which must have been let loose over the lunar surface each time a comet crashed in this manner.

The total amount of gas which can be acquired by the Moon in such catastrophic encounters is far from negligible. The average mass of a cometary nucleus is of the order of 10^{18} g; and comets that may have been ten or even a hundred times as massive are known from history (the most recent being comet Arend-Roland 1956h). The total mass of our terrestrial atmosphere (generated essentially by degassing of the Earth's interior) is known to be close to 5.1×10^{21} g, that is, about a thousand times that of an average comet; 3×10^{20} g would thus be sufficient to provide the Moon with a gaseous envelope containing about 1 per cent of the terrestrial air mass above each unit area of the surface; and a larger comet could bring proportionally more.

What is the chemical composition of gas that could be acquired by cometary impacts? An analysis of the spectra of cometary tails discloses the presence in them of a number of molecular constituents like C_2, C_3, CN, CO or CO_2 in neutral as well as ionized state. No one knows for sure the actual composition of the cold cometary nucleus, which is the sole source of gases evaporated from it by sunlight. Since, however, many of the observed gaseous constituents of cometary tails may have originated by photodissociation of more complicated parent molecules present in the nucleus, the latter may well contain constituents of molecular weight well in excess of fifty.

Now an appeal to the kinetic theory of neutral gases discloses that constituents of molecular weights close to twenty-five could remain gravitationally attached to the Moon for time intervals of the order of 10^3 years of daytime (and indefinitely longer at night), those of molecular weight of forty to fifty could remain so attached for 10^8 to 10^9 years! Time intervals of this order are comparable with the total age of the Moon, and certainly long in comparison

[22]

with an average time lapse between successive cometary impacts on its surface. Since, moreover, each such impact could have provided the Moon with an atmosphere of mass of the order of 1 per cent of the terrestrial air mass, giving rise to air pressure on the Moon of the order of a few bars, why is it not there?

The absence of any noticeable twilight phenomena on the Moon, referred to earlier in this chapter, leaves no room for doubt that the actual amount of gas around the Moon now must be less by several orders of magnitude to escape detection. On the other hand, it is equally certain that comets must occasionally strike the Moon; and the only way to reconcile this with the apparent absence of gas on the Moon now is to admit that cometary gas can be removed from the lunar environment faster than predicted by the kinetic theory of neutral gases, so that a complete dispersal occurs within time intervals which are short in comparison with the mean interval between cometary impacts.

Such a mechanism is indeed known to be available as soon as imported cometary gas gets ionized by energetic solar radiation, for as soon as this is accomplished, the electrostatic mechanism proposed by Öpik and Singer will take care of the removal of cometary gases as effectively as of any lunar indigenous gas. The speed of removal is, in effect, then identical with the speed with which the gas can be ionized, and depends essentially on gas transparency. If the respective air mass is small enough for the entire atmosphere to be optically thin in the ultraviolet range, a virtually total dispersal of such an atmosphere by electrostatic action can be accomplished in a time span of ten to a hundred days. If, on the other hand, the air mass is large enough to protect the bulk of its gas from hard radiation of the Sun by self-absorption, only in the outer fringe of semitransparence can gas be removed electrostatically, while dissipation of the rest follows the kinetic theory of neutral gases. Should this latter regime comprise the bulk of the atmospheric air mass, its mean lifetime could be increased from days to years and even to centuries.

This is, indeed, likely to be true after impacts of comets whose masses prove to be large. Even then, however, the mean lifetime of such an atmosphere would be fleetingly short in comparison with the age of the Moon, which is at least a few billion years old; consequently, the likelihood that we may find any gas left at any particular time is correspondingly minute. But, while it lasts, it may permit processes to occur which, even though short-lived, may leave a more permanent imprint on the stony face of the Moon for posterity to decipher. Thus, Harold C. Urey conjectured recently that certain formations on the Moon commonly called "rilles" (identification of which will be given in the next chapter) may indeed represent dried-up beds of fossil rivers, which may once have flowed over the respective parts of the lunar surface under the temporary protective umbrella of a transient atmosphere of cometary origin; but this view is as yet unproven and hypothetical. With comets around us in interplanetary space, it is thus impossible to assert categorically any longer that it has never rained on the Moon, or that its landscape was never swept by winds. We wish to reiterate, however, that with such evidence as we now possess, the role of these processes in shaping up the lunar surface—if not altogether negligible—must have been highly localized and very small indeed.

One last remark concerning the density of the present lunar exosphere may be added. We have seen earlier in this chapter that optical phenomena observable from the Earth rule out any gas envelope around the Moon of a ground density exceeding 10^{-12} g/cm³. On the other hand, the action of solar wind alone should be sufficient to endow the Moon with an exosphere of density of the order of 10^{-22} g/cm³ at the time of the "quiet sun," and as much as 10^{-20} g/cm³ during major flares. However, in recent years, we have come into possession of another tool of lunar research which could throw new observational light on the problem, namely, a whole class of artificial lunar satellites, the characteristics of which are listed in Table 3 of the Appendix. Collectively, these satellites completed more than 10^4 orbits around the Moon, of various eccentricity and size. If any appreciable atmosphere existed at an altitude of their periselenia passage (see column 4 of Table 3), a drag exerted by it on moving vehicles would have led to a secular loss of kinetic energy, manifesting itself through diminishing of the velocity of the spacecraft. A careful analysis of existing observations failed to detect any such secular change which could be indicative of air drag; and the upper limit above which period changes would have been detected relegates the present density of a hypothetical lunar atmosphere at 50 km aboveground to less than 10^{-15} g/cm³; but how far below this limit it actually is can still only be guessed.

If, therefore, the Moon possesses no detectable atmosphere its surface must thus be regarded as bone-dry; and to have been so, if not always, at least most of the time in the past. Thus, one of the most important agents causing geological changes on the Earth must have been largely absent on the Moon; and its action cannot be invoked in general to explain large-scale structural characteristics of the surface of our satellite, as documented by many photographs reproduced in this atlas.

A quest for other processes which could possibly do so more effectively must lead us to consider the state of the interior of the lunar globe containing the bulk of its mass; for (as is true of every celestial body be it a star, a planet or its satellite) this interior is the "engine room" which controls the large-scale structure and cosmic evolution of the respective body. The visible surface represents only the "boundary condition" of all thermal and stress processes going on in the interior—as well as an "impact counter" of external events which our satellite must have experienced since the days of its formation.

If we wish to examine the essential properties of the lunar interior, the primary clues are already in our hands, namely, the observed mass and the size of the Moon, combining as they do in a mean density of 3.34 g/cm³. What is the pressure prevailing inside this mass? It goes without saying that even in so small a celestial body existing under the influence of its own self-attraction, the internal pressure is essentially hydrostatic throughout most of the interior—meaning, in other words, that the strength of the material is unable to withstand its weight anywhere except, possibly, in the very outer part of its crust. If so, however, an elementary application of the theory of hydrostatics reveals that the internal pressure in a globe of the lunar mass and size cannot exceed some 50,000 atmospheres even at its center—a pressure exceeded at a mere 150 km below the surface of the Earth—regardless of the kind of material of which the Moon consists. Pressures of this order of magnitude are readily attained in terrestrial laboratories today; and the changes in density exhibited by common rocks under such pressures have already been measured. On the basis of all evidence we now possess, it is reasonable to conclude that the actual density of the lunar subsurface material is approximately equal to 3.28 g/cm³, and increases by compression to 3.41 g/cm³ near the Moon's center. Such a model fits in satisfactorily with the observed mean density of the lunar globe as a whole, and leads us to believe that the Moon consists of material which is very similar to that constituting the outer crust of our own planet.

Does the Moon represent, then, a solid and nearly homogeneous spherical rock, or is it partly molten in its interior? The answer to this question depends essentially on the sources of heat which would have been available for this purpose, and which until quite recently could only have been guessed at. Since 1966, however, considerable new light has been thrown on the problem of the actual rigidity of the lunar globe by contributions forthcoming from lunar-orbiting spacecraft (see Table 3 of the Appendix); and these have been so original as well as interesting that we must describe them in more than a few words.

2. The Moon—Our Nearest Celestial Neighbor

Perhaps the most interesting—because unexpected—was the discovery of relatively large and severely localized gravitational anomalies on the Moon, which accelerate the motions of overflying satellites. Such accelerations can be produced only by anomalous mass concentrations ("mascons," in brief) in the respective areas and none are apparent at the surface; moreover, the large observed rates of change in orbital motion indicate that the mascons responsible for them must be located at a shallow depth below the surface, and be relatively small in size (50 to 200 km).

P. M. Muller and W. L. Sjogren analyzed in 1968 the distribution of mascons over the lunar surface, and found them to coincide largely with the so-called "circular maria" on the Moon—a term that we shall describe more fully in the next chapter. Suppose—and this we shall also discuss in more detail in the next chapter—that these mascons represent leftovers of asteroidal bodies whose impacts created the circular maria, that is, cosmic "bullets" (possibly metallic) which hit the Moon in the distant past and became embedded in its crust at a shallow depth below the surface (comparable, in general, with the original dimensions of the impinging missile).

How long could these bullets remain embedded in a layer of finite rigidity defying persistent efforts of gravity to pull them down? A simple analysis of the mechanical aspects of the problem (carried out in 1969 by Urey and the present writer) discloses that if the Moon possessed the same rigidity as the terrestrial mantle, the mascons would be bound to sink to depths at which orbiting satellites could no longer sense them in a time of the order of 10 million years, that is, within less than 1 per cent of the probable age of the Moon. This explains, incidentally, why there are no mascons akin to the lunar ones on the Earth (which should have intercepted a comparable number of cosmic impacts, per unit area, as the Moon) at the present time; for they could linger near the surface for only a relatively short time.

Since, however, they appear to be present on the Moon, the only conclusion we can draw from this fact is an inference that *the lunar globe as a whole must be very much more rigid than the Earth;* in effect, about a thousand times more so if the age of the lunar mascons are to be of the order of 10^9 years. On the other hand, we know now (see Chapter 4) that the chemical (basaltic) composition of the lunar crust does *not* differ *grossly* from that of the terrestrial mantle. If so, the only way to endow our Moon with a rigidity sufficient to support its mascons for astronomically long intervals of time is to cool its material globally much below the level of temperatures encountered in the terrestrial mantle. In other words, *the crust of the Moon appears to be capable of tolerating much greater departures from hydrostatic equilibrium* than does our Earth; and this can be so if it is much cooler than the mantle of our own planet.

This result is, moreover, in agreement with recent measurements by the lunar Explorer 35 satellite (see again Table 3 of the Appendix) of magnetic interaction between the lunar globe and the solar wind—or, rather, the lack of one. For it appears that the Moon simply casts a geometrical shadow behind it in the solar plasma, thus behaving like an insulator rather than semiconductor; and its electrical conductivity (indicated by a lack of any interaction) should be less than 10^{-5} mhos/m. This can, however, be true for silicate material of which the Moon consists only if the *mean temperature of the lunar interior is less than about 1,000°C!* This is a much lower temperature than prevails in most of the terrestrial mantle. It is sufficiently low to endow the lunar globe with the requisite degree of rigidity, but gives very little encouragement to anyone expecting to find on the Moon any source of large-scale volcanism. The existence of local volcanic "pockets," to be sure, can never be ruled out by such "global" arguments as we have advanced so far. Nevertheless, such arguments lead one to expect that volcanic activity on the Moon—if any—should have been present on a much smaller scale than on the Earth.

Incidentally, the recent work of Explorer 35 provided us with one additional bit of information characterizing the physical properties of the lunar globe, namely, a well-nigh complete *absence of any magnetic field* of the Moon. That the magnetic field of the Moon—if any—is very weak was indicated earlier by experiments performed by the Russian Luna 2 in September 1959. As a result of more recent work of the American Explorer 35 we know now that the magnetic field of the Moon—again if any—does not exceed 10^{-5} gauss in strength, and that the magnetic moment of the lunar globe is less than one-millionth of that of the Earth to escape detection. This result confirms—if any further confirmation were needed—that the Moon does not possess any metallic core, and that such iron as may be present in the Moon's mass has not been thermally extracted from it.

3. The Lunar Landscape and Its Morphology

In the preceding chapter we became acquainted with the principal facts and figures which specify the fundamental properties of the lunar globe. With these in mind we now wish to invite readers to take a guided tour of the lunar surface, such as we would experience if we could approach the Moon by spacecraft and allow ourselves to be "captured" in a closed orbit around it. Because others—both inanimate and animate messengers from the Earth—have already done it, we can follow in their footsteps from the safety and comfort of our armchair at home. We shall not yet attempt to land, in our thoughts, on the lunar surface itself; this feat will be deferred till the next chapter. The present aim will be to furnish a largely descriptive survey of at least the principal types of formations which cover the lunar surface, the characteristics of which should help us to understand the nature of the most important processes which have been shaping the lunar landscape since time immemorial.

What is so arresting about this face of the Moon, and what can we learn from its inspection? From an analysis of the lunar rocks transported to the Earth by Apollo 11 we learned that between 3,000 and 4,000 million years have elapsed since the solidification of the pebbles picked up at its landing site in Mare Tranquillitatis; and the Moon has very probably been a close companion of the Earth during most of this time. A virtually complete absence of air or liquid water on its surface makes it certain, moreover, that most of its composite fossil record must be of very ancient age—its oldest visible landmarks being, perhaps, not far removed in time from the days of the origin of our whole solar system. On the Earth or other terrestrial planets, all landmarks of comparable age must have fallen prey to the joint disturbing action of air and water aeons ago. However, as any changes on the Moon—caused by other kinds of erosion—can proceed only at a very slow rate, its present wrinkled face must still bear at least traces of many events not far removed in time from the days of its formation.

The extent of the data now available for our inspection is indeed impressive. Of the 37.96 million km² of the lunar globe surface, more than 20 million km² of the front face (permanently visible from the Earth) have been observed and mapped by the methods of telescopic astronomy to a resolution between 0.5 and 1 km on the lunar surface. Moreover, since the advent of the spacecraft—and, in particular, of the U.S. Lunar Orbiters—the entire lunar globe (both its near and far side) has been photographically recorded to a ground resolution between 50 and 250 m on the Moon; and approximately 100,000 km² of its selected areas, to a resolution of 1 to 2 m. In fact, thanks to the Orbiters, we are now in possession of almost as complete a topographic record of the surface of the entire lunar globe as we possess for our own Earth; and the aim or this chapter is to survey briefly the salient features of this record.

Even to the naked eye the Moon is a beautiful object, diversified with markings which have been associated with numerous popular myths of many nations. If we look at its pockmarked face through a telescope (or, more comfortably, at many of the photographs reproduced in this volume), a cursory glance reveals the lunar surface to consist essentially of two principal types of terrain. One, rough and articulate, is comparatively light (reflecting, in places, as much as 18 per cent of incident sunlight) and broken up by many mountains. The other type is darker (reflecting on the average but 6 or 7 per cent of incident sunlight), much smoother and frequently so flat as to superficially simulate a liquid surface. The first type of ground is commonly called the continents. They occupy large continuous areas, particularly in the southern hemisphere of the Moon, and cover, on the whole, a little less than two-thirds of the area of the visible face of our satellite, though more than nine-tenths of the far side. The flatlands, or maria as they were misnamed by early observers of the Moon before the true nature of its surface was properly understood, checkered by rilles and wrinkle ridges, occupy the rest. Both types of ground are, on the whole, remarkably uniform in reflectivity and general appearance—on a small or large scale.

In observing the contrast between the two principal types of lunar surface on such photographs as reproduced on Plates 13–15, the reader should keep in mind that the difference in actual reflectivity of the continental and mare ground is actually much smaller than one would think. Photometric measurements of the albedo of the lunar surface (that is, the fraction of incident light back-scattered in all directions) vary from place to place only within the range 0.05 to 0.18 in visible light (becoming much less in ultraviolet, though greater again in infrared)—variations much smaller than we should encounter among common terrestrial rocks. The ratio of the reflectivity of the brightest and darkest spots on the Moon exceeds, therefore, scarcely a factor of three; while the continental areas are, on the average, not more than 1.8 times as bright as the maria.

A closer inspection of the lunar surface reveals a rather bewildering array of mountains and formations, no two of which are exactly alike. However, the dominant characteristic type of formation among them—and by far the most numerous on any part of the Moon—appears to be ringlike walled enclosures commonly called craters. This word is used here in its original sense to describe a cup-shaped topographic feature, as derived from the Greek root κρατερ (meaning "cup" or "bowl"), without prejudice for the views of their origin. Too specialized an interpretation could easily render it as much of a misnomer as the Martian "canals" or lunar "seas."

In order to facilitate references to the craters or the maria, the majority of more conspicuous formations of these types on the lunar surface have been given proper names assigned in accordance with certain general rules. Thus, most lunar maria traditionally have been given names connected with the weather: hence the etymology of Oceanus Procellarum (Ocean of Storms), Mare Imbrium (Sea of Rains), Mare Nubium (Sea of Clouds) or with a state of mind, such as Mare Tranquillitatis (Sea of Calm), Mare Serenitatis (Sea of Serenity) and Mare Crisium (Sea of Crises). Only certain small maria near the limb of the Moon were named after human beings—Mare Humboldtianum, Mare Smythii—or after their location—Mare Orientale; only one small mare on the lunar far side has been given a regional name, Mare Moscoviense—Sea of Moscow—by its Russian discoverers.

All major craters on the Moon traditionally have been given the names of persons no longer living, but who have entered the history of some branch of human endeavor. Some 500 such names are now recognized by the International Astronomical Union; and at present a similar system of nomencla-

3. The Lunar Landscape and Its Morphology

ture is being extended to the lunar far side. In earlier times, a veritable motley of personalities—saints and scholars, potentates, ancient sages and Christian ecclesiastics, but remarkably few artists or writers—found access to the lunar Pantheon by favor of their friends. Among the more recent candidates for immortality, potentates are now definitely out of the running, and the chances of the scientists have greatly improved—witness the first group of names bestowed on the formations on the far side of the Moon which now includes even a novelist, albeit one as closely associated with the Moon as Jules Verne. Lastly, lunar mountains are usually named after corresponding ranges on the Earth. Thus, on lunar maps, too, we encounter the Alps, Apennines, Carpathians, and the Pyrénées, whose names are familiar to the terrestrial geographer.

The largest complex of lunar sea beds is located on the front side of the Moon, in the form of the irregularly shaped Oceanus Procellarum, easily visible to the naked eye at the time of the full Moon. Several other—and not much smaller—maria appear to possess an oval (and roughly circular) form, bordered at least partly by mountain chains. The largest of these is Mare Imbrium (See Plates 16–24) and is also visible to the naked eye, and so are Mare Serenitatis, Mare Crisium, Mare Humorum and Mare Nectaris. All of these can be identified in the early drawings of Gilbert (Figure 1) or Harriot (Figure 3).

The lunar surface exhibits, however, examples of other and more enigmatic formations of this type. Perhaps the most striking one is the so-called Mare Orientale at the extreme western limb of the Moon, for a proper record of which we had to await the advent of the Orbiter photography (see Plates 45–47). This magnificent circular formation, consisting of a double ring of mountains with a sea bed at the center, is almost 900 km across and would be easily visible to the naked eye from the Earth if it were more favorably situated with respect to the line of sight. Only the eastern limb of this formation was glimpsed from the Earth before at a favorable libration, and given the name of the Cordillera Mountains.

Mare Orientale—the principal features of which are so strikingly illustrated by Plates 45–47—is, moreover, not unique on the Moon; formations (albeit smaller, or less well preserved) of similar form occur elsewhere on the lunar globe. Thus, Mare Nectaris on the visible side of the Moon (see Plates 99–102) is probably the "eye" of another formation akin to Mare Orientale, of which the Altai Mountains and the Pyrénées represent surviving parts of the walls. The whole formation is considerably less well preserved, but almost as large (790 km across) as Mare Orientale; and smaller formations of the same type are found repeatedly on the Moon's far side in a much better state of preservation (Plate 148 or 149).

Even a cursory glance at the enclosed fold-out maps of the two hemispheres of the lunar globe discloses the existence of another characteristic of the lunar surface which calls for explanation, namely, a conspicuous disparity in distribution of the sea bed between the two hemispheres. Ever since the first photographs of the Moon's far side were returned to the Earth, it has been evident that lunar maria are very largely situated on the hemisphere facing the Earth; and with a slight tilt of the principal axes of inertia of the lunar globe this asymmetry could be made almost complete.

What could be the cause of so striking an asymmetry? The clue was perhaps offered by another recent discovery, which we owe to the lunar-orbiting satellites of the last few years; and may be connected with the existence of discrete lumps of dense material (mascons) that were found in 1968 to produce local accelerations of mooncraft overflying them at low altitudes. The ways in which these mascons were located (and their bearing on the internal structure of the Moon) have been described briefly in the previous chapter. Here we wish to make only one additional point which may be connected with the asymmetric distribution of the maria over the lunar globe: namely, through the effect which these subsurface mascons are bound to exert on the moments of inertia about the principal axes of inertia of the lunar globe.

As the reader may be aware, the present Earth-Moon configuration—with the Moon rotating in the same period as it revolves—represents an approach to the state of minimum potential energy which the Earth-Moon system tends to attain in the course of time. It is possible (although it cannot as yet be proved) that the majority (if not all) of the circular maria do possess lumps of heavy matter at shallow subsurface depth. The presence of these lumps may (through their effect on the moments of inertia) orientate the lunar globe in such a way that the sum of the squares of the distances of the individual mascons from the line joining the centers of the Earth and the Moon be minimum; and most of the maria may have been orientated to face us in an effort to approach this state.

The dimensions of these circular maria are indeed huge, and make most of them easily visible by the naked eye from the distance of the Earth. Thus Mare Imbrium—the largest of them all—is approximately 1,000 km across; Mare Serenitatis, 610 km; Mare Crisium (Plate 63), 420 km; Mare Moscoviense (on the far side; Plates 135–136), 400 km; Mare Humorum (Plates 53–54), 320 km. Of the "double-walled" circular maria, Mare Orientale, the best preserved of this type (see again Plates 46 and 47) on the limb of the Moon's visible hemisphere, is almost 900 km in diameter; while a similar (though more dilapidated) formation on the visible side, centered on Mare Nectaris (see Plates 60 and 61) is only slightly smaller (some 790 km across).

Turning now to the craters—the second and most characteristic type of formation on the lunar surface—we note that they occur almost everywhere on the Moon, in continental regions as well as in the maria, no part of the surface being immune from these disfiguring pockmarks. Their numbers are, moreover, prodigious: those with diameters in excess of 1 km are estimated to add up to 300,000 on the visible hemisphere of the Moon alone, and to much more than 1 million on its far side; those smaller still are too numerous for a realistic estimate so far.

On the front side, the largest formations of this type visible through our telescopes attain dimensions well in excess of 200 km—such as the craters Clavius (230 km across; see Plate 77 or 78), Deslandres (240 km; Plate 59) and Janssen (238 km; Plate 61), the latter two being in a fairly dilapidated state. Even larger craters like Bailly (250 km), Pingré (270 km) and Schiller (280 km) are partly visible on the southwest limb of the Moon.

These dimensions are, however, far exceeded by craters discovered lately on the far side of the Moon. The largest of these—carrying the names of Hertzsprung, Korolev, and Apollo—are 530, 510, and 450 km, respectively; and others (as yet unnamed), almost equally large (see Plates 145 and 148) reproduce the features of, say, Mare Orientale so faithfully on a somewhat smaller scale as to suggest strongly a generic relation between them.

Altogether, there are nine craters on the Moon (all on its far side) exceeding 300 km in size; thirty-eight craters (thirty-three on the far side) whose dimensions exceed 200 km; and over a hundred (thirty-two visible from the Earth) with diameters between 100 and 200 km. Craters this size are usually characterized by fairly smooth floors, sometimes checkered by smaller craters (such as Clavius), which appear to be largely absent in others (for instance, Plato; see Plates 107–109). Of craters 50 to 100 km in diameter, typical examples are Copernicus, Theophilus, and Tycho, whose photographs from both ground-based and space-borne facilities are reproduced on Plates 78–79, 101–105, and 112–121. These craters are usually characterized by hummocky walls, rough floors, and a frequent presence of hills constituting "central mountains" (see, in particular, Plates 116–117 or 207). A hybrid type of crater exhibiting a central cliff rising out of a smooth floor is Tsiolkovsky (shown on Plates

139–142) on the far side of the Moon, attaining the dimensions of almost 200 km.

Craters substantially smaller in size exhibit again different morphological characteristics. Formations less than 10 km in size seldom, if ever, possess any central peaks; and their upturned lips are barely distinguishable from the surrounding landscape by their inclination. Finally, formations smaller than 1 km across—and these have been adequately explored only since the advent of lunar spacecraft—have hardly any outer walls at all; and constitute almost mere depressions in the ground which are not uniform in shape, and sometimes appear to be distinctly elongated (Plate 157). Moreover, far from possessing any central peaks, some may actually exhibit central depressions which have earned them the name of "dimple craters"—formations detected first on the televised pictures transmitted by the Rangers in 1964.

No two craters on the Moon are exactly alike. However, apart from the general distinguishing features which depend mainly on size, most such formations possess many characteristics in common. And as these are of obvious importance for any attempt to unravel with their aid past history of the lunar surface, which we shall attempt to outline in latter parts of this chapter, we shall describe them now in order to lay down the necessary background for subsequent interpretation.

First, *the distribution in size of lunar craters ranges continuously from the largest formations of this type to the smallest pits* discernible with optical means; craters in no size range appear to be missing. Second—and unlike the maria—*the craters appear to be distributed all over the Moon essentially at random* on a comparable type of ground, a fact suggesting that the events which produced them did not favor (or discriminate against) any particular part of the lunar globe. The reason why many more (and larger) craters are counted on the Moon's far side is a preponderance of the continental ground there (that is, the absence of the maria); but for equal areas of the same type of ground the number—as well as distribution—of all craters appears to be everywhere essentially the same.

Third, *the heights of the ramparts of lunar craters prove to be very small in comparison with their dimensions.* This is true, in general, for craters of all sizes when rampart heights are measured from the level of the surrounding landscape; though not necessarily for rampart heights above the crater floor. A difference between these (which, for small formations of this type, can be quite pronounced) bears out another fundamental characteristic of the lunar craters: namely, the fact that *their floors are generally depressed below the level of the surrounding landscape* by amounts which may, for large formations, run into several hundred meters. Moreover, the extent of this depression seems statistically correlated with the diameter of the respective formation—in the sense that the larger the crater, the deeper its floor lies below the neighboring landscape. Finally, the volume of the depression seems in many cases to come close to the volume of the ramparts raised around them, a feature suggestive of the possibility that the material contained in the rims may have been uplifted from the crust by forces which produced the entire formation. On the whole, craters on the Moon should thus hardly be considered real mountains in the terrestrial sense; but rather as "pockmarks" characteristic of the lunar surface and possessing no obvious terrestrial analogy.

Extensive measurements of the relative elevations and profiles of the ramparts of a great many craters from the length of the shadows cast on the surrounding lunar landscape in solar illumination (and, more recently, also by stereogrammetric work based on photographs taken from spacecraft operating in the proximity of the lunar surface) have revealed that, moreover, *the inward-sloping crater walls are much steeper than those sloping outward* which (especially for smaller formations) constitute but barely upturned tips. Thus, the maximum height attained by the rims of the large crater Clavius

(Plates 77 and 78) is only 1,600 m above the surrounding landscape, and not more than 4,900 m above the lowest point of that crater's floor.

For Copernicus, Theophilus, Tycho, and other craters whose photographs are reproduced on Plates 78–121, very much the same continues to be true. All three formations are close to 100 km in diameter, but the maximum altitude of the Copernican ramparts does not quite reach 1,000 m above the surrounding landscape, and 3,300 m above the crater's floor; the largest hill of its central mountain rises to barely more than 1,100 m above its surroundings. For Theophilus, the corresponding figures are 1,200 m and 4,400 m for the ramparts, and 2,200 m for the central peak; while for Tycho, these values proved to be 5,400 m and 1,600 m, respectively.

These results disclose that, in general, lunar craters represent surface depressions rather than any real mountains—as vividly documented by space photographs of Copernicus or Theophilus (see Plates 105 and 114). In particular, outward-sloping walls of most craters are inclined by not more than a few degrees to the horizontal; a contrary impression (formed in the past from cursory glances at ground-based photographs of regions close to the lunar terminator at the time of sunrise or sunset) disappears when the appropriate altitude of the Sun above the horizon (not readily apparent to inspection) is duly taken into account. Inward-sloping walls of many craters (especially of small ones) are often steeper and more rugged—witness high-resolution Orbiter photographs of a part of the inner walls of the craters Copernicus (Plate 115 or 119) and Aristarchus (Plates 125–126) showing vividly the broken structure of the respective type ground.

In the case of the pictorial evidence presented in this atlas, and of its descriptive principal features summarized so far in this chapter, what can we say about the *origin* of lunar relief, and, in particular, of its craters? A glance at the almost bewildering array of all sizes makes it perhaps unlikely that all of them originated in the same way, or at the same time; and a more detailed analysis of their features suggests that a suspicion of different origin is probably well-founded. In fact, the most reasonable approach to this problem can be made if we ask ourselves the following question: What are all the processes which could have conceivably cooperated in shaping up the surface of the Moon? And once we thus formulate our problem, we find ourselves facing two principal contending theories of crater origin, namely, the external theory—invoking the effects produced by *impacts* of other celestial bodies (asteroids, meteorites, or comets) on the lunar surface—and the alternative theory—relying on the *internal processes* connected with convection, gradual defluidization and degassing of the lunar globe consequent upon its build-up of internal heat due to spontaneous decay of radioactive elements, or any other activity which could be loosely termed "volcanic." In point of fact, *the entire surface of the Moon must be regarded as the outer "boundary condition" of all internal processes which may have been going on in the lunar interior since the origin of our satellite, as well as an "impact counter" of external events which may have visited it from outside.* In no other sense can an interpretation of the lunar surface possess any physical meaning.

That the surface of the Moon is full of inequalities was recorded by Galileo, the first telescopic observer of the Moon, in his *Sidereal Messenger* (1610); and this fact caused at first no small commotion among peripatetic philosophers of that time. Fifty-five years later, Robert Hooke, who, like Newton, was also interested in the Moon, dropped bullets into a pipe-clay and water mixture and saw formations arise which one could call "impact craters." But being an inquisitive soul, Hooke did not stop here but, as he tells us in his *Micrographia* (1665), he also boiled a mixture of powdered alabaster with water, and observed that this, too, produced transient craterlike structures on the surface of the liquid. Thus he started a very interesting controversy about the origin of lunar craters, which has not been definitely settled (at least in individual cases) ever since, in spite of a vast amount of new knowledge acquired in the last ten years by means of spacecraft, including Apollo.

3. The Lunar Landscape and Its Morphology

Instead, and partly as a result of this evidence, the two principal theories of the origin of the lunar surface features continue to face each other in contest for due recognition. In order to assess their relative merits and drawbacks as objectively as can be done at the present time, let us, in what follows, outline their main arguments and supporting evidence.

To begin with the theory of external impacts, we must remember that the interplanetary space through which the Earth, with the Moon, continues to circle around the Sun is not entirely empty. Far from it, for it contains a wide variety of ingredients of all weights and sizes: from the elementary particles (essentially hydrogen plasma) of solar wind, through microscopic specks of dust and larger meteorite debris (representing, probably, the leftovers from the time of formation of the whole solar system), to major meteorites, asteroids, or comets whose orbits through space may intersect the path of the Moon and occasionally collide with it. The frequency with which the surface of the Moon—like that of the Earth—undergoes direct hits by asteroids or comets is as yet difficult to assess with any accuracy—let alone what it was in more remote times. As, however, in the course of the long lunar past such hits must undoubtedly have been scored, it is important to realize the consequences which such events would entail; and these are indeed bound to be spectacular.

In order to visualize them at least to some extent, consider a moderately large meteorite—of the size and mass of a rock weighing one million tons ($m = 10^{12}$ g) and impinging on the lunar surface with a velocity v of, say, 30 km/sec—equal to that of the Earth in its relative orbit around the Sun. The kinetic energy $E = \frac{1}{2}mv^2$ of such a missile would be a quantity of the order of 10^{25} ergs and would enable it to penetrate into the lunar crust like a bullet to be buried well underneath the surface before coming to a complete stop. The kinetic energy which the meteorite possessed before impact must, however, be conserved, and its entire amount reappears in other guises to which it was converted in accordance with the accepted laws of physics—mainly as mechanical energy of shock and fracture, thermal energy and seismic energy of elastic waves. If, for the sake or argument, the entire kinetic energy of the meteorite were converted into heat, its temperature T would be given by $T = v^2/2C_v$, where C_v denotes the specific heat of the meteoritic material. As for stony meteorites, $C_v \sim 10^7$ ergs/g deg; this mechanism should generate a temperature T of the order of several million degrees, that is, sufficient to volatilize completely the whole impinging mass and convert it into an extremely hot bubble of gas imprisoned at a depth of several diameters of the original body beneath the lunar surface. In reality, the actual temperature of the impinging body would be considerably lower, because a large part of the original energy would be spent on other processes than heating; but still it is difficult to escape the conclusion that the impact of such a body would create local temperatures of the order of a few (possibly several) hundred thousands of degrees for a very short time.

Needless to say, such hot gas bubbles could not be contained by the weight of the overlying debris for time intervals longer than microseconds. They would immediately expand with great violence; and the effects of this expansion should severely affect regions that are very large in comparison with the size of the original missile. The main effect of the explosion should, therefore, be essentially that of a point-charge; and the initial direction of the impinging body could not have had much influence on the symmetry of the resultant surface markings. The probable result is schematically shown in Figure 7, which represents the expected cross-section of an impact crater. The amount of the actual solid material left around by the intruder should be negligible; most of it should have evaporated and escaped back into space, or become dispersed over a large part of the adjacent lunar surface.

Diverse terrestrial experiments with impacts of metallic or stony particles in brittle media—or with explosive charges ranging from microscopic dimensions to underground nuclear detonations in megaton range of explosive

[33]

power—are indeed found to produce local effects simulating the ramparts of certain types of lunar craters to at least a superficially astonishing degree (see Figure 8). Moreover, since the total energy requisite for producing terrestrial craters of given size in laboratory experiments, aerial bombardment of the ground or nuclear explosions is known, an extrapolation of this force should permit us at least to estimate the energies likely to be involved in the formation of much larger craters on the Moon by impacts of meteorites or other cosmic bodies.

In doing so we find that in order to produce a lunar crater 20 km in diameter by impact, a kinetic energy of the order of 10^{28} ergs would have to be expended in the effort; and to double or quadruple its size, energies ten or a hundred times as large would be prerequisite. The mass of such a body impinging with a velocity, say, of 20 km/sec would be 10^{16} g (10 billion tons) for a total energy of 10^{28} ergs, and proportionally larger for energies ten or a hundred times the amount considered. The diameters of solid spheres of such a mass (varying, in general, as the cube-root of the energy of impact) would, moreover, be approximately 1,200, 2,500 and 5,400 m if they were of stony material of average density 3 g/cm³, and 20 per cent smaller if their principal constituent were nickel-iron (of density close to 7 g/cm³). Thus, impact production of craters as large as Clavius or Deslandres would call for energies of the order of 10^{30} ergs (10 million megatons of TNT) associated with high-velocity impacts of small asteroids, of the size of Adonis, Hermes, or Eros (to name a few which paid rather close calls on the Moon in recent decades), whose dimensions are estimated to 10 km, and masses to 10^{13} tons.

Sudden expenditures of such prodigious amounts of energy would, however, trigger off a chain of events which would not only be devastating beyond imagination on and around the actual point of impact, but whose consequences should be felt, to a different degree, all over the Moon through the medium of seismic waves. In order to assess the magnitude of disturbances which can be caused by such waves, let us return to our impinging meteorite as it comes to a complete stop at some depth beneath the lunar surface—an event necessitating a conversion of its entire kinetic energy into other forms.

How will this large energy store be apportioned? Extensive work—both theoretical and experimental—discloses that, on the whole, only about 30 per cent of the total kinetic energy of the intruder will be converted into heat; 19 per cent into rock fragmentation; 50 per cent into kinetic energy of the ejecta; and about 1 per cent will be drained away from the focus of impact in the form of elastic (seismic) waves.

Thus we are led to conclude that a seismic energy of the order of 10^{26} ergs would emanate from an impact capable of producing a lunar crater 20 km in diameter, with proportionate modification for craters of other sizes; and the seismic waves excited by it should carry in their train the message of the event throughout the lunar globe to all parts of its surface. In other words, any impact of a body capable of producing a lunar crater would also be bound to set off a "moonquake," characterized by a very shallow epicenter; and the effects of such impacts on the Moon as a whole require some attention.

In order to appreciate more fully the significance of such moonquakes, let us recall that the most destructive earthquakes experienced on our planet

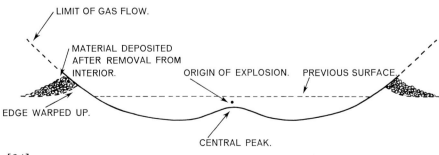

LIMIT OF GAS FLOW.

MATERIAL DEPOSITED AFTER REMOVAL FROM INTERIOR.

ORIGIN OF EXPLOSION.   PREVIOUS SURFACE.

EDGE WARPED UP.

CENTRAL PEAK.

7. *Schematic profile of a lunar crater created by impact.*

3. The Lunar Landscape and Its Morphology

within the memory of mankind entailed energy expenditures of the order of 10^{26} to 10^{27} ergs only, that is, 1,000 to 10,000 times less than that of a hypothetical moonquake accompanying the origin of a crater 100 to 200 km in diameter. When one considers the fact that there are over one hundred craters of this size on the Moon, and that the total number of those exceeding 1 km in size is more than a million, the cumulative effects of such disturbances over long intervals of time may produce effects which should influence the microscopic structure of the entire lunar surface; and in the next chapter we shall find that this is indeed so.

Individually this should be all the more true of larger (and more devastating) collisions of the Moon with planetesimals more massive than meteorites considered so far, which several investigators have invoked to account for the existence of the circular maria on the Moon—such as Mare Crisium, Mare Serenitatis, or even the largest of them such as Mare Imbrium or Mare Orientale—which may indeed represent huge impact craters produced by nearly grazing collisions of the lunar surface with low-velocity planetesimals in the early days of the existence of the solar system.

This view, first proposed by G. K. Gilbert in 1893, has more recently and more specifically been elaborated by that distinguished dean of lunar investigators of the present time—Professor Harold C. Urey. According to Urey, Mare Imbrium originated in the early days of lunar history (not far removed from the origin of the Moon as such), when a solid object about 200 km in size (about one-fifth of the dimensions of the resulting mare) and weighing some 5×10^{16} tons, made a nearly grazing plunge into the region

8. Nuclear explosion crater on the Earth (Sedan, in Nevada, USA) produced by the detonation of a shallow depth charge equivalent to (approximately) 100,000 tons of TNT (reproduced by courtesy of the U.S. Atomic Energy Commission).

of the present mare with a relative velocity of not more than 5–6 km/sec. The region of impact is thought to have been subsequently flooded with lava melted by heat liberated by impact. The velocity of impact, Urey observed, should have been less than the sound speed in the solid; for there is evidence (in the form of grooves seen on Plate 87 diverging radially from the region of impact) that the disintegration of the intruder commenced before the latter got completely buried underground; and the splinters of the impinging mass were flung away from the point of catastrophe to roll (rather than fly) over the lunar surface to a distance of several hundred kilometers, reaching places situated far below the horizon of the point of impact. The kinetic energy of the impinging planetesimal would have been of the order of 1.4×10^{33} ergs —equivalent to 30 billion megatons (3×10^{25} g) of TNT, or almost ten million times that of the most destructive earthquake ever experienced by man on our own planet. Moreover (according to recent work by van Dorn), Mare Orientale (see Plates 45–50) was produced by a more normal (that is, not grazing) impact of another planetesimal carrying a total kinetic energy of the order of 1.5×10^{31} ergs; and *mutatis mutandis* the same may have been true of all the formations of this type.

Gravitational anomalies discovered by Muller and Sjogren in the region of Mare Imbrium and other circular maria (Orientale, Serenitatis, Crisium, etc.), can be traced to anomalous subsurface mass concentrations (which could constitute the remnant of the mass of the cosmic intruder, or high-density modification of lunar rocks brought about by pressure generated on impact) and lend powerful—though not yet decisive—support to the view that circular maria on the Moon are nothing else but gigantic impact craters, differing from their lesser brethren in size rather than in kind.

Moreover, throughout our discussion we have so far been concerned with the effects wrought on the lunar surface by its collisions with *solid* bodies: meteorites, asteroids, or planetesimals. Any effort to understand more fully the principal features of the lunar face would be seriously incomplete without a parallel consideration of effects which could be produced by collisions with other known types of denizens of the interplanetary space: namely, the comets. According to the statistics available to astronomers at the present time, comets appear to be at least as frequent at our present distance from the Sun as are meteorites of comparable masses (that is, 10^{15} to 10^{18} g); and a wide distribution of the elements of cometary orbits is bound to render high-velocity collisions (in the range of 30–70 km/sec) with the Moon much more frequent than would be the case with the asteroids.

On the Earth, examples of both types of collisions and of their after-effects appear to be preserved, for, while the well-known Canyon Diablo crater at Canyon Diablo in Arizona was without doubt produced by the impact of a major meteorite in prehistoric times (and a few scores of other similar formations are now on record), the Siberian Tunguzka crater of 1908 was apparently produced (judging from a well-nigh complete absence of metallic components in the debris and other structural characteristics) by a collision of the Earth with a comet. For it is known that cometary heads—the only part of their structure which matters in the case of collision—represent but loose conglomerations of frozen hydrocarbons, with appreciable admixture of unstable chemical compounds (such as solid hydrogen peroxide, or azides), and these on impact would behave like high explosives, thus releasing chemical energy in addition to the kinetic energy of the head as a whole. Now, unlike a solid meteorite, cometary heads possess no tensile strength, and their impact on the Moon would scarcely indent the lunar surface to any appreciable extent. Instead, they would be completely volatilized at once and envelop for a short time the surrounding region in a stream of hot gas rapidly dispersing into vacuum.

Quite apart from the chemical binding energies, the kinetic energy of a major comet, such as Halley's, for instance, alone represents a quantity of the order of 10^{31} ergs; and the latter, if it could be converted totally into heat, would be equivalent to 2×10^{23} calories. If we assume, reasonably enough, that 2,000 calories are necessary to melt 1 g of lunar surface matter into fluid lava, a single cometary impact of this caliber could provide, for example, some 10^{20} g of lava capable of covering the 400,000 km^2 of Mare Imbrium to a uniform depth of some hundred meters. This may possibly furnish another explanation of the origin of lunar maria, compatible with the fact that no maria show any deformation in the regions where impact should have taken place and that their ramparts (as represented by the Alps and Apennine chains in the case of Mare Imbrium) are well below the horizon at the center of these great plains.

Comets with kinetic energies of the order of 10^{31} ergs are, to be sure, relatively rare. On the other hand, the number of plains on the Moon of the size of Mare Imbrium is also limited; and the probability that, in the past 4.5 billion years, the Moon may have suffered a considerable number of collisions with comets of requisite masses may be likely. Collisions with smaller comets may, in turn, have produced craters of the type of Archimedes or Plato, whose floors, surrounded by low ramparts, bear a striking similarity to the surrounding maria; but so far any such suggestion can be put forward only as a tentative possibility.

Having thus taken stock of the principal external agents, namely, impacts of various celestial bodies which can mutilate the lunar face over long intervals of time, let us turn next to examine the internal processes, whose action can affect the surface of the Moon in a similar manner. These latter processes are, in general, connected with the gradual build-up of internal heat by spontaneous radioactive decay possibly giving rise to slow convection in the outer layers. As a by-product of the secular heating (and aided by possible convection currents) the gasses and other volatile elements initially present in the interior of the lunar globe can make their way to cooler layers underneath its surface and accumulate there or escape into space. This defluidization and degassing, which may be partly operative on the Moon as it has been in the Earth, can—in its last stage—produce (by upwelling and withdrawal of a molten rock column) local areas of surface depression or subsidence, which the geologists on our Earth refer to as calderas.

Some terrestrial calderas bear indeed a rather striking similarity to many large lunar craters, whose most important morphological features—and this cannot be emphasized too strongly—are not their (often not too conspicuous) ramparts, but rather the general depressions of their floors *below the level of the surrounding landscape*. Thermal expansion and contraction phenomena accompanying defluidization could have brought about the occurrence of fracture patterns in the relatively solid lunar crust, which may have provided ducts of escape for the molten magma beneath.

Many typical mountain-walled plains on the Moon appear to be distinctly polygonal—in fact, hexagonal. Note, in particular, the group of large craters in the central parts of the Moon's apparent disk consisting of Ptolemaeus, Alphonsus, and Arzachel (Plates 87–89), Hipparchus and Albategnius (Plates 94 and 96–97); or Purbach, Regiomontanus, and Walter (Plate 86) to the south; and many others. Not only these, but even the largest formations of this type, such as Clavius near the south pole, or Mare Crisium on the Moon's eastern limb, show distinct hexagonal outlines when their foreshortening near the limb (as seen from the Earth) has been duly taken into account.

The hexagonal form by itself does not rule out impact origin. Witness the distinctly polygonal shape of the terrestrial meteor crater in Arizona (Figure 9) or of several major lunar craters like Copernicus, which are no doubt of impact origin. However, and this would be much more difficult to explain by impact hypothesis, many of the hexagonal craters (in particular, the one near the center of the Moon's apparent disk: see Plates 87–89) are similarly orientated and their sides more or less parallel, whether they are adjacent or separated by some distance. On the other hand, it is well known

[37]

that hexagonal shape is characteristic of the convection pattern which develops in a parallel layer of viscous liquid from below; and although the argument cannot as yet be definitely settled (on account of our incomplete knowledge of the distribution of heat sources in the lunar interior) it is open to surmise that some (possibly the largest) craters on the Moon may have originated as a result of internal processes, side by side with many others which are due to impacts; and while the number of the latter keeps accruing in the course of time, those which may have been produced by internal processes probably originated during one distinct epoch of the Moon's history and need not have been far removed from the time of its origin.

In the face of these and other facts, it is very difficult to escape the conclusion that both types of formative processes—external as well as internal—had a hand in shaping the face of the Moon as we see it today. Detailed analysis of all aspects of circumstantial evidence accessible so far from the Earth suggests that craters like Copernicus, Theophilus or Tycho are almost certainly due to impacts of solid bodies on the surface which must likewise have been solid down to a considerable depth. Such craters are relatively deep for their size, and are distinguished by their hummocky rims. The rims of Copernicus are closely simulated in structure by those of other craters equal or similar in size; and, in general, the ratio of the width of the hummocky terrain to the diameter of the crater diminishes with decreasing dimensions of the formation. Around some craters virtually all the rim terrain is made up of a nearly random arrangement of hummocks typical of the crest of the ramparts of Copernicus; around others the rim exhibits a pronounced radial pattern of low ridges typical of the periphery of Copernicus. The interior walls of these craters are invariably terraced (see Plate 119 for Copernicus, Plate 125 for Aristarchus) and quite steep; the floors are rough (see Plate 81 for Tycho); and nearly all possess a single or multiple central peak.

However, perhaps the most revealing identification of primary impact formations are the families of secondary craters and (at least for younger formations) the bright ray systems diverging radially from them. The process

9. *Terrestrial meteor crater near Canyon Diablo, Arizona, USA, due to the impact of a large meteorite ten to fifty thousand years ago.*

3. The Lunar Landscape and Its Morphology

by which a lunar crater can originate by an impact of a cosmic intruder was outlined earlier in this chapter; and its essential feature was found to be the ejection of a large amount of solid debris by the shock generated on impact; the mass distribution of this debris may range from fine dust to rocks of size limited only by their internal strength. Some of this matter may be ejected with velocities sufficient for escape from the lunar gravitational field; but most of it will fall back on the Moon, at a distance depending on the initial velocity and angle of ejection. On impact, such rocks or boulders are, in turn, bound to produce "secondary craters" with velocities less than that of a circular orbit around the Moon (1.68 km/sec; otherwise they would not have fallen down)—perhaps around 1 km/sec on the average. If so, however, the secondary particles (thrown out from the lunar surface, and falling down upon it) should possess about a thousand times less kinetic energy per unit mass than the primary intruder, and this fact should leave also a distinct mark on the type of surface formation resulting on impact: namely, secondary craters should not only be much smaller, but also shallower than craters due to "primary" cosmic impacts, in which the impinging body buried itself much deeper; and their walls should be lower, giving them a characteristic washed-out appearance.

The clustering of not hundreds, but thousands of such secondary craters around well-known formations like Copernicus (Plate 112), Theophilus (Plate 101), and Langrenus (Plate 62) discloses beyond doubt their primary impact origin. There are, however, other and much larger craters on the Moon, such as Clavius, Ptolemaeus, or Alphonsus, where such origin is doubtful; for they lack most of the distinct characteristics (hummocky walls or floors, central peaks) of impact formations; and, most revealingly, the attendant hosts of secondary craters are completely absent from their surroundings. It is true that their polygonal outlines cannot, by themselves, be invoked to rule out impact origin. What are, however, the chances that two neighboring craters of distinctly hexagonal outline like Ptolemaeus and Alphonsus (Plates 87–89) would fit in neatly together as is observed, *sharing one confluent side*, if they were produced by successive impacts? Their formation must obviously have been simultaneous; but even if—a most improbable event—two large meteorites struck the Moon at exactly the same time and so close to each other as to give rise to Ptolemaeus and Alphonsus, what are the chances that the ramparts raised in this way would share one side, undamaged by destructive interference, which we clearly see at work in the impact pair of Theophilus and Cyrillus on Plates 103 and 104?

Consider, moreover, details of the ramparts of the crater Alphonsus, as revealed by the recent photographs of Ranger 9 (Plates 91–93) in March 1965. From the vantage point of approaching spacecraft, the ramparts of Alphonsus proved to consist of a rather loose system of ridges, the gentle slopes of which show evidence of much less celestial bombardment than the rugged and darkened floor of the crater, suggesting that the oldest landmarks visible on the crater floor are much older than the walls.

Moreover, in a few places patches of flat ground can be seen in between the individual rampart ridges and—interestingly—their surface appears to be just as dark and heavily pockmarked as that of the main crater floor (see Plate 92). This suggests that the flat patches in between the ramparts are generically related with the floor, and may once have been a part of it before the ramparts were raised. Moreover, the fact that the surface structure of patches, later blocked off by ramparts, was not altered appreciably by this event suggests that the raising of the ramparts was a gradual, rather than cataclysmic, process; for the latter—such as an impact—would have certainly buried preexisting surface structure under a thick layer of debris.

This argument strengthens our belief that the crater Alphonsus (and, by implication, other formations akin to it) is not of external, but internal, origin. Its floor represents probably the oldest type of lunar surface, containing a continuous record of events going back to the oldest chapters of the

[39]

geological history of our satellite. Its ramparts are probably younger—but not as young as the surface of the adjacent plains of Mare Nubium to the west; for the latter are at a higher level than the crater floor; and their crater density is very much smaller (see Plate 91). It must have been the walls around the crater that protected its floor from being submerged by whatever covered up the mare later; and the different crater densities on the respective types of surface, together with their differences in altitude, permit us to unravel their stratification.

Another large crater on the Moon, which could scarcely have been produced by impact, is the enigmatic Wargentin near the western limb of the Moon (see Plate 84). It is more than 86 km across—therefore, almost as large as Copernicus or Theophilus. However, its floor—instead of being depressed below the level of the surrounding landscape—has been raised to the level of its rims; thus giving the impression of a crater whose bowl has been "filled to the brim." Wargentin is, moreover, no longer unique on the Moon in this respect; for another (smaller) formation of this type has recently been discovered by the writer in Orbiter 5 photographs of regions near the Moon's north pole (see Plate 75). Neither of these two formations could have been produced by impact, and their existence suggests the action of some kind of uplift; but the mechanism of such an uplift remains still largely obscure.

The escape of volatiles by degassing of the interior constitutes an inevitable, long drawn-out cosmic process, which is bound to occur in any solid body radioactively heated from within but should not be identified too closely with the volcanic processes as we know them on the Earth. The latter represent essentially superficial phenomena, of much smaller order of magnitude and lateral extent. Yet volcanic activity on Earth follows, in general, as an aftermath of defluidization; and should the same be true on the Moon as well, it may have left its imprint inside the depressions of lunar calderas as it did in their presumed terrestrial homologues. These are likely to be much less conspicuous than the calderas themselves by size, but still demonstrable by their particular characteristics.

In order to identify at least one such formation on the lunar surface, let us inspect an Orbiter photograph of the Moon reproduced on Plate 86. Within the walls—possibly a caldera—called Regiomontanus we find a small "hilltop" crater usually designated as Regiomontanus A (marked on Plate 86 with a white arrow). It represents a small cone, whose hill rises barely 700 m above its surroundings; and the crater on its top is but little more than 5 km across (see the accompanying Figure 10). By its cross section, Regiomontanus A resembles a terrestrial volcano of the Krakatoan type, but—and this is essential—it is also unique; no other formations of this type have so far been found anywhere else on the Moon.

And this gives rise to the question of where all the other volcanic cones are on the Moon. Formations like the extinct volcanoes of the American Pacific Northwest—Mount Baker, Glacier Peak, Mount Rainier, Mount Adams, Mount Shasta—or Mount Fujiyama in Japan would be easy to detect on the Moon if they were there; but, instead, they appear to be conspicuous by their absence. The only inference we can draw from this is to realize that volcanic activity as we know it on the Earth must have been largely absent on the Moon or must have occurred on a much smaller scale. This conclusion is what we should really expect if the interior of the Moon is as cool and rigid as we

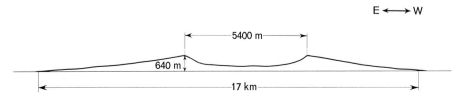

10. East-west profile of the "hill-top crater" Regiomontanus A (marked on Plate 86 with a white arrow), the external characteristics of which recall volcanic calderas on the Earth.

[40]

3. The Lunar Landscape and Its Morphology

surmised in the last chapter from different lines of evidence; and the absence on the Moon of volcanic cones of the shape or size of Mount Rainier or Fujiyama constitutes a further independent support of the view that the Moon has in the past been volcanically much less active than our Earth.

So far throughout this chapter we have been concerned with lunar craters and maria, which certainly represent the most typical morphologic features of the surface of our satellite. Besides these, the surface of the Moon exhibits also several other characteristic features of lesser magnitude possessing no obvious terrestrial homologues, four of which—the domes, rilles, wrinkle ridges, and bright rays—deserve special mention. The lunar domes represent small, inconspicuous hills of an approximately circular base and gently convex cross section, clustering in certain regions of the lunar surface in considerable numbers. The largest formations of this type attain linear dimensions of 10 to 12 km and altitudes of 300 to 400 m; but the majority are smaller; and the inclinations of their slopes seldom exceed a few degrees. Large groups of individual domes of this type occur in Oceanus Procellarum, especially in the region of the Marius Hills (see Plates 26–27); on the eastern shores of Mare Tranquillitatis (Plate 64 or 159); or on the floors of certain large craters like Copernicus (Plate 121).

Already the best terrestrial photographs of such formations (as reproduced, for instance, on Plate 159) disclosed indications of craterlike central depressions on the top of some domes; a feature fully confirmed by high-resolution photographs (such as the one reproduced on Plate 160) taken from lunar-orbiting spacecraft. On the whole, the domes occur almost exclusively in the plains of the maria, or in flat floors of large craters; and preferably in groups. In other words, they seem to grow in a certain type of ground; and their origin is undoubtedly internal; but the actual mechanism of their growth is still largely obscure.

The wrinkle ridges—another characteristic feature of lunar maria—are possibly three-dimensional (that is, elongated) versions of lunar domes. They represent low ridges running like shallow submerged veins underneath the surface of the vast plains of lunar maria; and meandering sometimes for several hundred kilometers. Typical examples of such ridges near the crater Posidonius photographed from the Earth are shown on Plates 67 and 69, while Plates 28–37 show a view of similar ridges in the vast plains of the Oceanus Procellarum photographed by spacecraft from much closer proximity to the lunar surface.

Topographic studies of the profiles of such wrinkle ridges show that the inclinations of their slopes are only of the order of 1 to 2°. As such, they can be distinguished much more easily from above (especially at the time of sunrise or sunset) than on the ground itself; and astronauts on the Moon may find it hard to recognize one when they come across it. The origin of the wrinkle ridges is again indubitably internal, and possibly connected with processes that gave rise to the domes.

The rilles represent an altogether different type of lunar surface formation, without any obvious terrestrial analogy. They scarcely constitute any homogeneous group on the Moon, either. For instance, the conspicuous rille in the neighborhood of the crater Cauchy (Plates 159 and 160) represents a shallow valley with a nearly level floor, depressed below the level of the surrounding flatlands by only a few percentages of its width. Formations of this type south of the craters Sabine and Ritter, in Mare Tranquillitatis, recorded by Ranger 8 from close to the lunar surface, are shown on Plate 155. Others— typified by the Hyginus or Ariadeus rilles in the Mare Vaporum —are deeper and apparently represent surface scars less completely healed. The well-known Schröter Canyon near the twin craters Aristarchus and Herodotus (Plate 124) belongs to a similar category. Still other types of lunar rilles are narrower and deeper meandering gorges, like the well-known Hadley rille paralleling a part of the Apennine shores of Mare Imbrium, a similar rille running through the middle of the Alpine valley (Plate 168), or around the crater Prinz. Such rilles are also known to occur on crater floors, where they may run roughly parallel with the walls (such as they do, for instance, in the crater Alphonsus), but also almost perpendicular to them (in Lacus Mortis, for instance; see Plates 71–72).

The origin of the rilles of all types must be sought in the stresses and strains of the underlying ground and is, therefore, internal. Some—especially the shallow rilles as shown on Plates 171 and 173—are probably subsidence formations (trenches?); but the cause of subsidence remains as yet unknown. Some of the deeper rilles—whose depth is comparable with their width—may, possibly, represent dried-up canyons of fossil rivers, which may once have temporarily run over the respective parts of the lunar surface under the protective umbrella of a transient atmosphere. But, needless to say, all this is as yet purely speculative, and without any independent confirmation.

Most rilles described above and illustrated in the atlas occur in the maria. Related with these on the continental surface are grooves of the type illustrated on Plates 87 and 149, often cutting across other mountainous formations on their way. Whether these are due (like rilles) to subsidence, or to ricocheting effects of solid debris ejected by major meteor impacts, remains as yet uncertain. Another type of lunar formation is clefts—such as the Straight Wall in Mare Nubium, which separates for a distance of about 60 km two parts of this mare differing in altitude by about 300 m. The slope of the dividing wall (determined again from the shadow phenomena) appears to be less than 11° over the entire length of the wall, and the cause of a subsidence of one part relative to the other is as yet unknown, though the cause was undoubtedly tectonic. Photographs of this formation taken at sunrise as well as at sunset are reproduced on Plates 57–59.

Lastly, several craters on the Moon—though not a very great number—are surrounded by systems of bright rays, resembling the ejecta patterns around terrestrial explosion craters. Bright rays associated with the craters Tycho, Copernicus, and Kepler represent perhaps the most conspicuous examples of such systems, but several other systems can be discerned elsewhere on the Moon as well. One of the most brilliant systems of bright rays has been photographed by the Apollo 8 mission over the Moon's far side emerging from the crater Giordano Bruno (see Plate 144). Some, like those diverging from the crater Proclus on the western shores of Mare Crisium (see Plate 63), spread over only a little more than half a circumference; while others like Messier (Plate 62 or 129) possess only two parallel rays reminiscent of a cometary tail.

Large ray systems are seldom characterized by complete symmetry and may (like Tycho's) include rays which do not intersect in the common focus. The most extensive lunar system of rays is that diverging from the crater Tycho; and some of its rays—like the one traversing Mare Serenitatis—exceed 2,000 km in length. The ray system associated with the crater Copernicus extends over 500 km from its center, but all other patterns of this type are smaller.

The rays accompanying lunar craters consist essentially of loop-shaped streaks of brighter material than that of the maria over which they have been splashed. By their reflectivity, the rays appear to be an extension of the crater rims and cannot be sharply distinguished from them. Moreover, their major arcs and loops can be often locally resolved into a system of feather-shaped elements, ranging between 15 to 20 km in length, radially spreading out from the crater (see Plates 188 and 189).

The rays do not exhibit any measurable vertical relief (that is, they cast no shadows), and the variation of light reflected from their surface in the course of the lunar day suggests that even their microrelief is essentially the same as that of the surrounding darker landscape, from which they differ mainly by their relatively high albedo. They represent probably nothing more than thin layers of ejecta from an impact crater thrown out by the initial explosion, and deposited all around in ballistic trajectories.

[42]

3. The Lunar Landscape and Its Morphology

In concluding our brief guided tour covering the principal types of formation characteristic of the lunar surface, let us stress with equal emphasis other types of formations, familiar to us on the Earth, which we did *not* find among them. The entire surface of our satellite shows no indication of any mountains or mountan chains which could have been formed by the *folding* of the crust —a process so familiar to the geologist on Earth—nor of any trace of formations that could be due to lateral motions of the Moon's crust. Such mountain chains as we find on the Moon border mostly large maria, and may constitute nothing but partly destroyed ramparts of gigantic impact craters which were formed early in the history of the lunar surface. Thus, the chains **that we call** lunar Apennines, Caucasus, Alps, Juras, and (possibly) Carpathians may all be parts of a continuous wall that once enclosed the entire Mare Imbrium; while the Haemus and Taurus mountains (together with the Alps and the Apennines) did the same for Mare Serenitatis. The Altai mountains and the Pyrénées may constitute the last extant parts of a continuous mountain chain that may once have surrounded the entire Mare Nectaris, just as the Cordilleras encircle a part of the Mare Orientale. Thus folding—the most important orogenic process continuously active on the surface of our own planet —seems to be completely absent on the Moon, just as the principal types of formations apparent on the lunar surface lack any obvious terrestrial homologues.

That this should be so is only natural, and due in effect to the great disparity in mass between our own planet and its only natural satellite. In the preceding chapter we saw that the smallness of the Moon's mass has deprived it of the possibility of generating or retaining (except, perhaps, for very short intervals of time) any air around it, or water on its surface. A complete absence of any effects that could be produced by these agents is obvious (on any scale); and accounts for the fundamental difference between the external appearance of the Earth and the Moon. Also a globe as small as the Moon can generate and retain but little internal heat; and thus is condemned to remain cooler in its interior than a planet of terrestrial size. In consequence, the crust of the Moon can be expected on general grounds to be solid rather than plastic down to a much greater depth than the Earth's is and the large-scale structure of the lunar surface bears this out by a total lack of any evidence of folding, or any lateral crust motions which on the Earth would be called "continental drift." The bulk of the Moon's mass is apparently too cold (and, consequently, too solid) to allow any appreciable differential motions to arise in its globe, or, at any rate, near its surface; and the morphology of its entire surface—based now not only on telescopic evidence, but also on all information furnished by spacecraft—is entirely consistent with this picture.

In the present chapter we have limited ourselves so far to a descriptive survey of the principal types of formations encountered on the lunar surface, together with a brief account of the processes by which such formations could have originated. We now raise—and attempt to answer—the following question: *Can the different types of formations encountered on the lunar surface be arranged in some kind of relative stratigraphic time sequence?* While absolute calibration of any such sequence will have to await determinations of the radioactive ages of rocks belonging to the different strata, the construction of a relative stratigraphic sequence of events on the Moon can be attempted now by at least three independent (though indirect) methods of approach, namely, (a) the principle of overlap; (b) the degree of ruggedness of the surface; (c) the extent of ground reflectivity.

Of the three, the principle of overlap is perhaps the most obvious and dependable. If there are two craters that overlap each other—and the photographs reproduced in this atlas show many examples of such a situation—then the one with the unbroken rim must be more recent than the one whose rim was damaged or entirely removed. In the case of two overlapping craters of comparable dimensions, such as the pair of Theophilus and Cyrillus on Plates

[43]

103 and 104, an application of this principle requires Theophilus to have been formed after Cyrillus. But the principle can be applied also to a situation in which a large crater contains smaller ones within it (for an illustrative example, see the photograph of the crater Clavius reproduced on Plates 76 and 77). No process that raised the ramparts of Clavius would have left the smaller formations seen undisturbed within it. Therefore, small craters now seen on the floor of Clavius must represent creations subsequent to that of the large configuration. On the basis of this evidence Clavius is, therefore, considerably older than all other craters which can at present be seen on its floor. The greater the number of small craters inside a large one, the greater should be the disparity between their ages.

Crater overlaps of great multiplicity can be found in certain parts of the lunar surface; and in some places it is possible to arrange thus five or six craters in a time sequence. Another relative age criterion (supplementing overlap) are the streaks of relatively bright material (the rays) which diverge in all directions from certain impact craters. As these rays must have originated (by ejection) at the same time as the parent crater, the latter must obviously be younger than any feature overlayed by its rays. Such rays represent, indeed, tentacles spreading widely over certain parts of the lunar surface and enabling us to extend the system of relative dating as far as they reach.

The main importance of the preceding age criteria lies, however, in their application to the dating of the maria. If we accept the foregoing premises, there seems to be no escape from the conclusion that the oldest parts of the visible lunar surface are those which are most rugged, and contain the greatest number of craters or other types of mountains. For no matter whether the operation of either the external or internal processes of crater formation on the Moon has been uniform, or diminished with time, the oldest parts if its surface should obviously have accumulated the greatest number of scars. If this is indeed so, then the oldest parts of the lunar surface are to be sought on the far side of the Moon, or, on the near side, in continental regions surrounding the south pole. Regions which appear to be equally old can, of course, be found also elsewhere on the Moon on a smaller scale; see, for example, a photograph of the regions near the Moon's north pole as reproduced on Plate 75; or the floor of the crater Alphonsus as seen on Plate 93. Such regions may indeed contain in their sculpture an unbroken record of events of the very distant past—going back, as we have said, possibly to times not far removed from the origin of the solar system. But as the mean density of craters in most of the great dark plains of the lunar surface is much less than that encountered near the south pole, it follows that the maria should be younger than the mountainous continental regions, and some of the great craters like Copernicus, Kepler, and Aristarchus, which spread the tentacles of their bright rays over large parts of the surrounding maria, must be younger still.

In point of fact, the brighter any element of the lunar surface, the more recent its present relief is likely to be, for the gradual infall of interplanetary dust alone is bound to darken the ground and lessen the differences of its reflectivity in the course of time. This process has, in turn, an interesting application to the chronology of impact craters. As we have mentioned, many craters of this group are the foci of prominent ray patterns, but others are entirely unaccompanied by rays. Thus, Eratosthenes (Plate 110) is a good example of a crater that exhibits all the principal topographic features of Copernicus and is surrounded by a well-developed pattern of gouges, but completely lacks rays. Moreover, where not overlapped by the Copernican ejecta, the walls as well as the floor of Eratosthenes exhibit relatively low reflectivity.

But—returning to the maria—as long as we are ready to accept as a working hypothesis that the dominant process responsible for shaping lunar relief is cratering by external impacts (which, coming as they do from random directions, do not favor any particular part of the surface), the *mare plains must be younger than the continental regions*, because fewer hits have been

[44]

3. The Lunar Landscape and Its Morphology

scored there. It is impossible as yet to translate crater density into a time interval, for the intensity of cosmic bombardment may have been very different in the past from what it is now; and a large part of the lunar sculpture visible now may have been chiseled in the earliest part of the history of our satellite. But there are other witnesses to be seen on the mare plains to show that this surface may once have been more heavily cratered than it is now; these are the so-called "ghost craters," or rings.

Ghost craters—such as are clearly seen on several plates included in this atlas—represent little else but the rims of the ramparts of ancient craters which were later flooded by deposits (lava?) covering the surface of the maria. Perhaps the largest of them are the Flamsteed ring (Plate 132) in Oceanus Procellarum, Stadius (Plate 110) in Sinus Aestuum, and Daguerre (Plate 101) in Mare Nectaris, or a beautiful ghost preserved on the northern slopes of the crater Aristillus (Plates 19 and 20) in Mare Imbrium. Incidentally, these ghost craters provide us with unexpected means of gauging the actual depth of the overlying strata from the visible dimensions of protruding rings. Earlier in this chapter we mentioned that the diameters of typical craters are statistically correlated with the heights of their ramparts. If we now assume that the submerged craters conformed to the same relation at the time of their formation, a difference between the present height of their ramparts and that corresponding to their visible dimensions would represent the depth of the overlying strata. This depth ranges, in general, from a few hundred meters up to one or two kilometers in the Imbrian system —that is, it is very shallow in comparison with the lateral extent of these strata.

The actual age of the entire lunar surface will, we repeat, be established only by absolute dating of the respective strata by radioactive methods. The view prevalent among students of this subject assumes the Moon to have been formed as we see it today about 4.6 billion years ago, by an accumulation of solid particles of small sizes at relatively low temperature, decidedly lower than required for volatilization of the respective material. On the other hand, the radioactive age of rocks recently brought home by the Apollo 11 astronauts from the surface of the lunar Mare Tranquillitatis was determined to be between 3 and 4.6 billion years (see Chapter 4). This strengthens our present belief that the stony structure of the lunar surface as we see it today, including the maria, was essentially completed in the first 500 to 1,000 million years of lunar existence; and has since been subjected by nature to only minor retouching.

4. Surface of the Moon and Its Structure

In the preceding chapter we completed our bird's-eye tour of the lunar surface as seen from above at various heights, and considered briefly the meaning of what we saw. The aim of this chapter will be to descend figuratively to the surface itself—in the wake of the spacecraft that preceded us—and with the aid of the results brought (or sent) back by them to try to understand the nature of the environment in which the first travelers to the Moon found themselves on July 20, 1969 as they set foot on the ground of this alien world.

Until the advent of lunar spacecraft in the last decade, our knowledge of the structure of the lunar surface remained limited to what could be learned about it from indirect evidence. The smallest details discernible on the Moon with terrestrial telescopes are hardly ever less than 400 to 500 meters in size, not because the resolving power of instruments could do no better (the theoretical resolution limit of a perfect 200-inch telescope should be close to 80 meters on the Moon in visible light); but because large optical surfaces are never perfect, and—above all—because the quality of the telescopic images is seriously degraded by the unsteadiness of atmospheric seeing at almost all times and all places. In point of fact, the atmosphere overhead but seldom allows us to attain the full resolving power of a perfect 40-inch aperture almost anywhere in the world, let alone of larger telescopes; and details on the Moon less than half a kilometer in size are invariably lost in a haze of light diffraction, unsteadiness of atmospheric "seeing" and (for photographic work) the irregularities in film grain.

Nevertheless, certain definite facts about the structure of the lunar surface were known before the advent of lunar spacecraft, from observations that could be made from the Earth with telescopes of moderate size—nay, with the naked eye. Perhaps the one most directly obtainable is the "light curve" of the Moon, representing the variation of its total brightness with phase in the course of each month. The most astonishing feature of this curve is its steep slope: a rapid rise of light toward full Moon, and an equally rapid diminution after the full phase has been passed. The full Moon is only twice as large in illuminated area as the Moon at the time of the first or last quarter; yet photometric measurements disclose the full Moon to be nineteen times as bright as the time of first quarter. Moreover, there appears to be a real surge in intensity of moonlight, more than doubling its brightness within only hours just before zero phase, and losing it with equal speed after full phase has been passed (the "opposition effect").

This rapidity of light changes before and after the full Moon is quite inexplicable in terms of the diffuse scattering of light on a smooth surface of any composition and has caused our astronomical predecessors to conclude that the *structure of the lunar surface must be highly broken and vesicular, and the surface itself replete with innumerable pits which begin to cast appreciable shadows almost as soon as the Sun has ceased to stand directly overhead.* In other words, a rapid loss of brightness of sunlit landscape on the Moon in the course of an afternoon can scarcely be due to anything else but *shadows* cast on itself by its own irregularities. This was the conclusion arrived at from photometric studies by the majority of astronomers of at least twenty to thirty years before the advent of spacecraft which at last opened this surface structure to a more direct inspection (see Plates 190–201); and the reader can judge for himself the correctness with which we have anticipated the outcome.

These findings are also entirely consistent with the fact—already noticed by Galileo in the early part of the seventeenth century—that the apparent brightness of every element of the surface attains a maximum at full Moon, regardless of its relative position on the lunar disk, be it a part of the continents or the maria. The apparent disk of the Moon indeed exhibits no "limb-darkening"; and the photometric homogeneity of its face, affirmed with considerable accuracy by more modern observations, suggests also the cause of the underlying surface roughness which gave rise to its honeycomb structure: for *what else but an external influence, such as continued infall of micrometeorites* (against which the surface is in no way protected by nature) *and "cratering" produced by them could impress the same uniform kind of microrelief on any type of surface all over the Moon?*

All these conclusions (which anticipated correctly the results disclosed in 1966 by soft-landing mooncraft) have been deduced from observed photometric properties of the Moon in visible light, which represent nothing but sunlight incident on the Moon and scattered by its surface in the direction of the Earth. In Chapter 1 we mentioned that the Moon is, on the whole, a rather poor reflector; for only a little more than 7 per cent of light incident upon it is scattered in all directions. What happens to the rest? The balance of it must obviously be *absorbed* and used up in *heating* the surface. But any body possessing a finite temperature must also *emit* radiation of its own, and its characteristics should depend essentially on its absolute temperature. For our Sun, whose effective temperature is a little more than $5,700°K$, most of its radiation is emitted at optical frequencies—between the violet and the red end of the visible spectrum (with a maximum in the yellow)—the whole giving an impression of what we call the "white light." This is also the color of reflected moonlight; but the lunar radiation proper is of a very different kind.

The sole and sufficient reason for this difference is the fact that the temperature of the lunar surface is so much lower than that of the Sun. The Moon, like the Earth, receives all its heat by radiation from the Sun; but because their average distance from the Sun amounts to 214 solar radii, each square centimeter of lunar surface receives only about one in 214^2 or one 46,000th part of the heat flux passing through each square centimeter of the surface of the Sun. As this flux (for a black body) is known to be proportional to the fourth power of the absolute temperature, it follows that the mean temperature of the Moon should be $\sqrt{214}$ or about fifteen times lower than that of the Sun—approximately $380°K$ (that is, about $+107°C$). This supposes, of course, that the Sun is standing in the zenith, with its light falling normally on the illuminated surface; should it fall obliquely, the radiation would be diluted, and the temperature maintained by it would be lower.

These elementary considerations lead us to expect that the temperature on the Moon will not be more than $380°K$; and if so, its radiation should be very different from white light—most of it being emitted in the deep infrared —with the maximum around the wavelength of 10μ (microns) or 0.01 mm. Light of this color is, of course, quite invisible to the human eye and incapable of impressing an ordinary photographic plate. It would, in addition, experience considerable difficulty in penetrating through our terrestrial atmosphere in between the interlocking absorption bands of water vapor and carbon dioxide. However, that part of it which does pierce through can be detected and, in fact, measured quite accurately by its thermoelectric effect.

Such measurements, performed extensively in the last ten years, have indicated that at the subsolar point in the lunar tropics—when the Sun stands directly overhead—temperatures as high as 130°C are reached each day. During the afternoon—as the zenith distance of the Sun increases—the temperature steadily declines to well below freezing at the time of sunset and continues to decline further in the course of the night until an appalling minimum of between −180°C and −190°C is reached before sunrise. The total range of day-to-night variation in temperature on the Moon is, therefore, little more than 300°C, and ranges from the temperature above that of boiling water at noontime down to that of liquid air at dawn. This is, of course, the maximum range encountered in lunar tropics. Near the poles, where the Sun never rises high above the horizon or sets completely for very long, the temperature variations become correspondingly smaller; but even there the range is rather frightening and apt to dampen the enthusiasm of many a would-be space explorer!

Why does the Moon behave so differently from the Earth in this respect? For both the Earth and the Moon the principal source of heat is, of course, sunlight; as our planet and its satellite are, on the average, equally far from the Sun, both are bound to receive equal amounts. But their particular properties allow them to husband this energy in essentially different ways. The visible light of our satellite, we repeat, represents only a few percentages of the total energy actually received from the Sun. The bulk of incident sunlight is absorbed by the lunar surface and reradiated as infrared light—far too red to be visible to the human eye, but easily measurable using its thermoelectric effect. The Earth does the same; but its own absorbing and protective mechanism is completely different from anything existing on the Moon.

From the global climatic point of view, by far the most important reason for this difference is the fact that approximately three-fifths of the Earth's solid surface is covered by oceans; and their water is a substance of very remarkable physical properties. For our globe as a whole, the oceans can absorb a great deal of heat from the Sun without becoming too hot and can lose much of this heat without becoming too cold. Moreover, through the action of ocean currents, the heat and cold can be distributed over thousands of miles to the terrestrial surface by mechanical action.

And this is not all, for the continents and oceans are surrounded by a gaseous atmosphere. This atmosphere is almost completely transparent to the bulk of the white light of the Sun; but two of its minor constituents: namely, water vapor and carbon dioxide—constituting together but 0.1 per cent of the entire atmospheric mass—are almost completely opaque to the infrared radiation given off by the Earth and its oceans. The bulk of the heat sent out by a black body of approximately 300°K absolute temperature is effectively trapped by the absorption of CO_2 and H_2O in our atmosphere which acts like a greenhouse, thus making its own contribution to the mitigation of extreme temperatures encountered on the Earth.

How different all this is on the Moon! The absence of any atmosphere precludes the existence of any liquid over the free surface (which would evaporate and dissipate almost immediately even during cold spells of the night), and solid rocks possess a heat capacity which is indeed very small in comparison with that of water. As a result, their surface can rapidly become hot during the daytime and cold again at night; and because there is no means of storing heat or cold, the extremes are great, though the mean temperature of the Moon is not so very different from that of the Earth.

In one respect, however, the lunar climatic changes are not so drastic as a first glance at the foregoing figures may indicate. Although the extremes are great, the duration of the lunar day—lasting 27.32 days or 656 hours of our time—is again so long that the rate of change of the temperature is not actually so impressive: a drop from 110°C to −180°C in 14¾ days separating the lunar noon and midnight corresponds to a mean temperature gradient of less than 1°C/hr, which is certainly not so remarkable. It is its persistence over so many hours of rise and fall which makes the extremes in temperature so far apart. To be sure, much more rapid temperature changes occur on the Moon about once a year (once every twelfth lunar "day") during the relatively brief intervals of lunar eclipses, when the Moon passes through the shadow cast by the Earth into space. Such eclipses last only two to three hours of our time; but while they last, the Moon experiences almost as large a change in climate as it does between day and night. In particular, the egress of the Moon from the shadow is accompanied by a steep rise in temperature of almost 200°C in less than one hour, or more than 3°C/min. Even so steep a temperature gradient should not—we wish to stress—cause any thermal cracking of the rocks, which could contribute to their disintegration; mechanical action of impacts is vastly more effective in achieving this.

If the wide range of temperatures on the Moon should, perchance, give some less courageous individuals second thoughts about visits to such inhospitable places, let us reassure them quickly that these large climatic changes appear to be limited strictly to the exposed surface, and become very much reduced immediately beneath it. How do we know this? From measurements of the thermal emission of the lunar globe in the microwave domain of the spectrum, that is, at wavelengths longer than 1 mm, which were opened up by observers from the Earth in the last twenty years, the results were truly revealing. First, the range of temperature variations deduced from the measured intensities of thermal emission of the lunar globe in the domain of radio-frequencies (at millimeter to meter wavelengths) proved to be less than those measured in infrared light, becoming the more reduced the longer the wavelength. Secondly, the maxima and minima of the temperatures deduced from the microwave measurements were not shown to follow the altitude of the Sun above the horizon, but *lagged behind* the surface temperatures by a phase-shift that increased with the wavelength.

How to account for such phenomena? Why, in particular, does the amplitude of diurnal temperature changes diminish so rapidly with increasing wavelength, and which reason causes the diurnal heat wave to lag increasingly in phase behind the surface insolation? The basic clue is the well-known fact that radiation observed at different frequencies does not originate at the same depth but is coming from layers which, in general, are located more deeply the longer the wavelength. In other words, the surface of the Moon—opaque to visible and infrared light—becomes partially transparent in the domain of microwaves; and the lower their frequency, the deeper inside we penetrate with their aid. In practice, a limit to this prospecting in depth will be imposed by the fact that the thermal emission of the Moon diminishes rather rapidly with increasing wavelength in the microwave domain, and the lunar radiation at wavelengths in excess of 1 m cannot be separated from the instrumental sky background noise. In the millimeter to decimeter wavelength range the readings are, however, significant and enable us to penetrate approximately down to 1 m beneath the surface in tracing the flow of heat due to diurnal insolation.

If the variation of microwave temperatures is damped as rapidly with increasing depth, and continues to lag in phase behind the infrared (surface) temperatures as the observations appear to indicate, the only reason for it can obviously be a very low thermal conductivity of the lunar surface layers. The observations definitely indicate that the daily variation in temperature at a depth of 30 cm already amounts to less than one-third of its surface range, and the effects of diurnal heat waves do not make themselves felt there till after a time lag of some eighty hours. Moreover, at a depth of about one meter any variation completely disappears and a constant temperature of about −35°C obtains day or night. These data should make it possible for us to evaluate the coefficient of heat conduction of the lunar surface material, sufficient to retard the flow and attenuate it as is observed.

4. Surface of the Moon and Its Structure

When such computations were made for the first time, investigators received a considerable jolt, namely, the coefficient of heat conduction required to account for these phenomena turned out to be about a thousand times smaller than for any known terrestrial rocks. This was so, however, only as long as we compared the computed lunar conductivity with that of solid rocks, in which heat flows through their entire body. In a broken material, on the other hand, heat can propagate only by flow through the actual areas of contact between the individual grains or pebbles—areas which can be diminished by loose packing. When the computed coefficients of lunar heat conduction were compared with those of loosely packed powders of common terrestrial rocks, some disparity still persisted, but it was almost completely removed when the laboratory measurements were repeated in a vacuum (otherwise heat would be propagated also through air between the dust grains).

Thus, while photometric observations of the Moon in visible light produced adequate evidence of the fact that its surface must be rough and porous, observations of lunar thermal radiation in the microwave domain of its spectrum disclosed that the broken structure of the surface must extend to at least a foot or so under the visible surface. This we knew years before the first soft lander—and, eventually, men—probed the surface on the spot. But even these findings were exceeded in depth by what we learned in the last quarter of a century by another method of active exploration of the Moon at a distance—by means of radar echoes.

In Chapter 1 we listed as the commencement of the lunar space age the day of September 13, 1959, when the first particles of terrestrial matter in the form of the Russian spacecraft Luna 2 crash-landed on the surface of our satellite. Should we, however, wish to date the beginning of the space age from the day when the Moon was reached by electromagnetic radiation sent out from the Earth, it would be January 10, 1946—another memorable landmark in the history of the subject—when scientists of the U.S. Army Signal Corps in Belmar, New Jersey, beamed toward the Moon a series of short radar pulses emitted by a mere 3-kw transmitter operating at 110 Mc/sec, and—behold—approximately 2.56 seconds later echoes were recorded, demonstrating that within this interval the signals sent out from Belmar completed a round trip of half a million miles to the Moon and back. The real triumph of this experimental technique was to amplify the echoes sufficiently to make them audible, for although the energy fed into the antenna of the transmitter was 3 kw, the returning echoes carried an energy of little more than 10^{-17} w.

This first radar contact with the Moon was repeated a month later from Hungary and has never been lost since that time. In the last 25 years it has provided us with a wide range of valuable scientific data, of which only one aspect, directly relevant to the structure of the lunar surface in depth, will be mentioned here: namely, the unexpected weakness of the observed radar echoes from the Moon.

Ever since the first echoes of radar pulses sent out to the Moon were recorded in 1946, it has been found that these echoes were considerably weaker than they should have been for a globe of lunar distance and size consisting of solid rocks. In order to reconcile theory with the observations it was necessary to assume that the effective dielectric constant of the lunar surface material was about three times smaller than that of common silicate rocks. Between 1946 and 1966 the same conclusion was arrived at by countless experiments with radar pulses at very different wavelengths which were reflected from surface layers of very different depth (equal, in most cases, to several multiples of the wavelength). It was not until pulses at decameter wavelengths were first reflected from the Moon in 1964 that the strength of the echoes began to point to dielectric constants approaching those of solid rock. These echoes (for pulses corresponding to wavelengths of 15 to 19 m) were returned from subsurface depths equal to several times their wavelength.

On the basis of the mean density of 3.34 g/cm³ of the lunar globe, it has long been surmised that the Moon consists essentially of silicate rocks similar to those common in the Earth's mantle; and recent α-scattering experiments performed in different parts of the lunar surface by Surveyors 5 to 7 in 1967–68, as well as the results of the Apollo missions in 1969, have not only confirmed this surmise, but have narrowed down its uncertainty to rocks of composition probably akin to that of the terrestrial basalts. Under these conditions, the low effective dielectric constants inferred for the lunar surface from radar echoes could be reconciled with the basaltic composition only by low volumetric concentration of the material, that is, by assuming that the reflecting layers are not solid, but consist of an accumulation of rubble and loose debris with a considerable volume of empty space between individual grains or stones. This is the same conclusion which had been previously arrived at from studies of thermal-conduction properties of the crust of the lunar surface and, more recently, by direct sampling of the lunar surface material by the soft landers.

However, none of these latter methods possess the penetrating power of radar echoes at decameter wavelengths; and these have indicated that diminished volumetric concentration of lunar surface material extends down to a depth of 50 to 100 m—a long way beyond the reach of any sampling device now in operation or contemplated. As the radar echoes emerging at decameter wavelengths possess low angular resolution (with the optics employed, the Moon would in their light appear as a single picture point), low volumetric concentration of the surface material deduced from them to a depth of 50 to 100 m represents only a global average which can be exceeded in some places and not attained in others. However, that the Moon borders on space through a brecciated lithosphere, a shell of broken material (of relatively low volumetric concentration) many meters in depth, can now be accepted, not only as a theoretical expectation, but also as an observed fact and as a global characteristic of the Moon to which the geologists refer as the "regolith."

The cause of this mechanical damage, namely, primary cosmic bombardment, is, to be sure, not operative on the Moon alone, but also on any other celestial body equally exposed to such bombardment. In particular, our Earth must have received a comparable bombardment, per unit area, in the same length of time; since for cosmic bodies causing greatest mechanical damage —intruders with masses greater than 10^6 g—our atmosphere provides next to no protection. However, the combined action of air and water would obliterate any wounds caused by cosmic bombardment very much more rapidly than these are inflicted, so that our lithosphere shows no evidence of a global fragmented layer. The fact that our Moon does exhibit evidence of such a layer discloses that the "healing agents"—air and water—are not only absent on the Moon now (as we know abundantly also from other sources), but that they probably were of no importance, even temporarily, over the long astronomical past of our satellite.

Thus, as a result of this work, we are now led to picture the outermost crust of the Moon as consisting of a stony regolith produced by cosmic abrasion of the surface of the Moon through external impacts. Its extent in depth may vary from place to place, but its global average is unlikely to be less than several dozen meters. It is also clear to us now that a cosmic body of the size of the Moon (or, for that matter, of Mercury, and to some extent, of Mars) must interface with outer space through the layer of such a regolith; the latter represents as natural a boundary condition for a globe of the lunar size as the hydrosphere and atmosphere do for planets like the Earth or Venus.

The question can, however, be asked: Are there any other formations or phenomena actually seen on the lunar surface which would bear out the existence of the regolith? A quest for such indications should lead us to an inspection of the structure of the surface—and, in particular, of the maria—on the 1 to 10 m scale, which was discovered on photographs taken by high-

4. Surface of the Moon and Its Structure

resolution lenses of the lunar orbiting satellites between 1966 and 1968. In order to introduce our evidence, attention is invited to Plates 116 and 207 featuring the well-known lunar crater Copernicus as recorded by the Orbiter moderate-resolution and high-resolution photography. Plate 116 merely introduces the location; but our subject really commences with Plate 207, which reproduces a high-resolution photograph taken by Orbiter 5 of the slopes of one of the hills constituting the "central mountain" of this crater. The ground resolution of Plate 207 is between 1 and 2 m on the lunar surface.

The arresting feature of Plate 207—like that of Plate 117 or 127—is the presence of a great many boulders—mostly 2 to 20 m in size—on the slopes of the central hill and their accumulation around the foot of the mountain, in contrast with their scarcity on the flat crater floor. This is, moreover, no specific characteristic of this particular region; wherever the Orbiter photography recorded for us inclined slopes on the Moon at a sufficient resolution, such as the floor of a part of Bode's rille shown on Plate 173, boulders of comparable size are seen to have detached themselves from the inclined slopes of the rille and rolled down to its bottom. In soft ground, such rolling boulders are apt to leave snakelike tracks behind them to indicate the direction of motion (see Plate 205 or 206); or distinct traces of avalanches down the inward-sloping crater walls can be identified, such as the one shown on Plate 126 in the crater Aristarchus, leaving no room for doubt as to the mechanism which produced them. Boulders of the size shown in the accompanying photographs (and doubtless many more below their limit of resolution) constitute a natural ingredient of the lunar surface (or, at least, of its outer layer, down to a depth of 50 to 100 m). In this surface they may be intermingled with finer-grain soil, from which they can be pried loose by mechanical action (moonquakes?) and, propelled by gravity, roll down a slope into the valley.

Preliminary statistics of size-frequency distribution of lunar boulders, based on high-resolution Orbiter photographs of different parts of the lunar surface, have disclosed that a large majority of individual boulders of this type is less than 10 m in size, and very seldom—if ever—exceeds 20 m. The well-known "spires" photographed by Orbiter 2 on November 7, 1966 in Mare Vaporum (see Plate 202) constitute boulders the largest of which is about 15 by 20 m in size—its height above ground having been determined from the length of its shadow cast on the surrounding landscape, while the Sun was approximately 11° above the horizon.

The fairly well-defined upper limit on the size of the individual stray boulders is probably imposed by the limited strength of the constituent rocks; for larger rocks could probably not have been made, or transported, in one piece. The limited size makes it also obvious why their discovery in large numbers had to await the advent of Orbiter telephotography; for none of them would have been visible on plates taken at more moderate resolution. The views televised from the proximity of the soft-landing Surveyors did, of course, detect exposed boulders on the lunar surface some months before the advent of the Orbiters (see Plate 212) and enabled us (in the immediate proximity of the soft lander) to trace the size-frequency distribution of lunar stones down to the centimeter size. In one respect, however, the evidence provided by the Orbiters constitutes a greater challenge to students of the lunar environment, namely, that the larger size of the Orbiter-detected boulders makes their presence on the lunar surface all the more thought-provoking, and their ubiquitous presence compels us to recognize them as landmarks as typical of the Moon on a 1 to 10 m scale as lunar craters are on a larger scale.

High-resolution photographs of the floor of the crater Copernicus have disclosed other interesting and characteristic features. In order to demonstrate these, we reproduce on Plates 208–210 a group of small domes that abound on this part of the lunar surface—small hills 200 to 300 m across and generally less than 40 to 50 m in height. The characteristic feature of

[53]

these domes is, however, the fields of stony "gendarmes" covering the brows of these hills, giving them the characteristics of "hedgehogs." More than thirty such hedgehogs can be located on the floor of Copernicus, one of which is shown in detail on Plate 210. Their characteristics are all alike: they constitute boulder fields on the tops of low domes, clustering where surface curvature is smallest, and are well-nigh absent on the slopes or near the foot of these hills as well as on the flatlands between them.

The significance of such formations may be accounted for essentially by arguments relying on the role of landslides propelled by gravity. As in the case of the larger hills shown on Plates 117 and 207, the material of the small domes consists of a mixture of boulders with smaller debris. The stony gendarmes on top of the hedgehog-like domes probably represent boulders more securely perched on shallow hills to withstand processes which may send smaller debris intermingled with them sliding downhill. Thus they become denuded in time except for the "gendarmes."

The explanations advanced for all phenomena described thus far possess several features in common. They all postulate the boulders to be an essential ingredient of the lunar ground material in each particular region, which can roll downhill when pried loose by mechanical action of yet unspecified kind. This explanation fails conspicuously, however, to account for the presence of boulders and boulder fields which have been discovered by the Orbiters (and Surveyors) to occur also in complete flatlands of the lunar surface—boulders which under no condition could possibly have rolled down from anywhere in the vicinity; and it is to these much more enigmatic stray boulders that we now turn our attention.

In order to illustrate the predicament in which we find ourselves in the case of boulders freely exposed to view in the lunar flatlands, let us look at a section on Plate 211 (approximately half a kilometer square) of lunar surface in the region of Mare Tranquillitatis, secured with the high-resolution lens of Orbiter 2 on November 7, 1966. The photograph discloses the presence of boulder fields of quasi-circular pattern (reminiscent somewhat of the terrestrial "megalithic rings"), the individual members of which are of the order of 10 m or less in size. Another photograph reproduced on Plate 212 and televised to the Earth in June 1966 by Surveyor 1 from Oceanus Procellarum illustrates another boulder formation, similar to those features on Plate 211 but smaller (the individual boulders here being feet rather than yards in size), demonstrating again that the features under discussion are not exceptional but apparently common on the Moon. How to account, however, for their existence and exposed positions?

The first part of the problem facing us in this connection is one of origin. Where could such boulders have come from? Since there are, in the present cases, no hills in the neighborhood from which rocks could have rolled, the only other possibility which comes to mind is impact. The possibility that any stone visible on the lunar surface on Plates 211 and 212 could represent a primary impact of a stony meteorite from interplanetary space can, of course, be dismissed almost immediately, for such bodies are known to strike the Moon with velocities of many kilometers per second; and no stone impinging on solid surface with such speed would have the least chance to come to rest on the exposed lunar surface for us to see it. Such cosmic intruders would get buried underground at depths equal to many times their original dimensions, where they would probably be vaporized (by a conversion of a large part of their initial kinetic energy into heat), or at least completely shattered into small fragments.

It is known, however, that all major primary impacts capable of giving rise to impact craters of the size of Copernicus or Tycho eject in this process 10^2 to 10^3 km^3 of lunar material, a large part of which will fall back on the lunar surface almost all over the Moon, along suborbital trajectories characterized by terminal velocities of the order of 1 km/sec. Local explosions caused by primary impacts can eject—and accelerate to suborbital speeds—

[54]

4. Surface of the Moon and Its Structure

boulders of the size we see on the lunar surface on Plates 211 and 212. But, and this is essential, they cannot land them gently enough to survive impact in one piece, and in exposed positions as we see them today.

In order to demonstrate this in more specific terms, consider the bearing strength of the material of the lunar surface as revealed in 1966–67 by the soft-landing Surveyors. The area of each one of the three footpads on which these spacecraft came to rest was close to 700 cm^2, supporting at touchdown a weight * of approximately 300 kg. Therefore, for a total support area of 3×700 cm^2, the spacecraft would have exerted on the lunar surface a static load of 3×10^5 g$:3 \times 700$ cm^2 = 143 g/cm^2. In actual fact, the Surveyors landed on the Moon with a terminal velocity close to 3½ m/sec, and the dynamical load at touchdown was sufficient to depress the ground by (typically) about 10 cm below the level of the adjacent surface.

Now a cubic rock of side a and density ϱ will weight ϱa^3, exerting a static load per unit area equal to ϱa. For silicate (basaltic) rocks of composition similar to that indicated by the α-scattering experiments of Surveyors 5 to 7, the density ϱ should be close to 3 g/cm^3. If so, however, a cubic basaltic rock should exert the same static pressure of 143 g/cm^2 as the Surveyor's legs if its side $a = 48$ cm. And should this rock have landed on the lunar surface with the terminal velocity of the Surveyor (3 to 4 m/sec), it should have penetrated it to a depth equal to 21 per cent of its side.

The dynamical load exerted per unit area is known to increase (for moderate loads) approximately with the square of the velocity of impact. If so, however, the rock would have buried itself completely underground for impact speed around 10 m/sec; for velocities of the order of 100–1,000 m/sec (typical of secondary or tertiary impacts) the impinging missiles would come to rest underground at a depth equal to many times their actual dimensions. This depth could, of course, be lessened if the target area were to consist of harder ground than that on which the Surveyors or Lunas landed in the maria. But in such a case no solid rock could withstand the impact shock without being shattered into pieces too small to be resolved by the telephoto lenses of the Orbiters; thus the photographs taken by them would confront us with no problem.

In other words, according to any plausible theory which could be advanced, such rocks or boulders as the reader sees on photographs reproduced on Plates 211 and 212 should really not be there! For boulders occurring in flat marial regions of the lunar surface, far from any conspicuous mountains, could scarcely have arrived at their present locations otherwise than along some ballistic trajectory by impact. If the target ground were soft (as indicated by all the soft landers), the rock should have buried itself underground well out of sight of the external observer; if, perchance, the target ground were hard, no silicate rocks moving at 100 to 1,000 m/sec could have survived the impact as boulders 1 to 10 m in size.

Is there a way out of this perplexing dilemma? In an effort to find it, let us recapitulate the facts bearing on the case which appear to be observationally well-founded, or at least probable on grounds of more circumstantial evidence:

(a) Boulders 1–10 m in size are shown by Orbiter photographs to occur individually or in groups on many parts of the lunar surface which are far from any cliffs or conspicuous mountains; these boulders consist probably of basaltic material of density close to 3 g/cm^3.

(b) In the absence of any mountainous ground in the vicinity, these boulders have probably arrived at their present localities by ejection from other parts of the Moon, along suborbital trajectories characterized by secondary impact velocities of the order of 100–1,000 m/sec.

(c) The target surface possesses a bearing strength of the order of 10^5–10^6 dynes/cm^2, as indicated by the soft landers, down to a depth of at least

---

* In the terrestrial gravity field.

[55]

a few meters (far beyond the reach of the instrumentation of the soft landers); for otherwise impacting basaltic boulders would get shattered into pieces too small to be resolvable on the Orbiter photographs.

(d) If soft ground is to cushion the impacts so as to enable rocks 1–10 m in size to survive in one piece, these rocks must be able to penetrate into the ground by at least a few times their original size. In other words, rocks transported by secondary impacts should disappear out of sight in the maria, and only shallow craters or depressions should mark their burial place immediately after the event.

(e) If, therefore, such rocks eventually emerge again on the surface to become visible to an external observer, it follows that they must have subsequently been *exhumed* by some kind of erosive process that gradually removes the debris in which the rocks originally became embedded. This seems to be the only possibility for large rocks to be seen now in one piece in the midst of the relatively soft surface far away from any exposed cliffs or mountains.

What kind of erosion could be invoked to accomplish this task? It goes without saying that such processes as micrometeoritic bombardment or sputtering caused by the solar wind, which endow the lunar microrelief with its peculiar photometric properties, are completely ineffective in this connection. However, another process is known to exist which can better accomplish the purpose, and this is the secular cumulative effect of moonquakes.

That seismic quakes must occur on the Moon not infrequently (astronomically speaking) is without doubt: for quite apart from possible internal seismicity of the lunar globe—the extent of which is still hypothetical and difficult to estimate—the body of the Moon is bound to suffer a quake each time a cosmic collision with any asteroid, meteorite, or comet gives rise to a primary impact crater.

One consequence entailed by this intermittent but long-drawn-out cosmic bombardment of the Moon by external bodies must have been a gradual break-up and shattering of its surface rocks (through the action of shock and seismic waves emanating from shallow epicenters of individual impact points) down to a considerable depth. That this has been so is no mere theoretical surmise, but an actual fact discovered by an analysis of the radar echoes observed from the lunar surface discussed earlier, and also, more recently, by the response of the Apollo 11 seismometer operating on the Moon since July 1969, the output of which shows evidence of microseismic activity in the lunar regolith.

Another agent which can help to sort out lunar debris by size is recurrent moonquakes triggered by tides that are raised in the Moon's globe twice a month by the terrestrial attraction. The seismometers left on the Moon by the Apollo 12 mission have registered such tidal effects to recur with clock-like regularity; and there is no doubt that, in the long run, such effects contribute to the litho-exhumation of the coarse material in the lunar regolith.

But having made a case for a global layer of fragmented material to extend to a depth of many meters over most of the lunar surface, let us inquire whether or not the existence of such a layer can help us to understand the presence of boulders in exposed regions of the lunar flatlands, as shown on Plates 211 and 212. The answer is indeed in the affirmative; for if such boulders form a constituent part of the lunar regolith, having been embedded in it by secondary impacts (or produced by shattering effects of primary impacts *in situ*), they could be made to surface by cumulative shake-up action caused by individual moonquakes of external (impact) as well as internal origin. For it is well known from soil mechanics that one of the most effective ways to sort out loose debris or gravelly material of the same density, but differing in size, is to administer to the mixture a mechanical shake-up, in the course of which large particles will eventually emerge on the surface, being supported by smaller particles which tend to slip through and subside toward the bottom.

[56]

4. Surface of the Moon and Its Structure

The presence of individual boulders on the level parts of the lunar surface—in exposed positions which the static bearing strength of the surface is barely able to support—demonstrates the outcome of a gradual mechanical "erosion" in the course of which *boulders embedded in loose regolith are gradually lifted ("litho-exhumed") by shake-ups produced by moonquakes or landslides.* According to this view, the megalithic rings of boulders shown on Plate 211 represent the debris of a much larger rock which once fell there and split up underground into pieces; what we see now are remnants of this event—in the form of fractured remains—which have gradually made their way to the surface by litho-exhumation.

An almost continuous record of microseismic activity, telemetered down to the Earth since July 1969 from the seismometer that was installed on the Moon as part of the Apollo 11 mission, has already demonstrated that processes which could cause litho-exhumation are indeed operative in the lunar crust now, and may have been since time immemorial. Moreover, a piece of iron—obviously of meteoritic origin—was picked up from the exposed lunar surface by the magnets aboard Surveyor 7 in 1968. Since this iron could have reached the Moon only by a primary impact with a velocity of at least several kilometers per second, it must have initially penetrated to a considerable depth. If it was found now resting on a relatively smooth surface, this proves that it must have been lifted there from its initial burial depth by the same process which also lifted up the boulders.

Having described thus the principal structural characteristic of the lunar landscape, as disclosed by spacecraft of the last few years, let us turn our attention to the chemical and petrographic composition of the lunar rocks and surface debris as we see them on photographs reproduced on Plates 200–212. Until the advent of the last three soft-landing Surveyors 5, 6, and 7, the composition of the material constituting the Moon could be surmised only from indirect evidence, such as the mean density (3.3 g/cm^3) of the lunar globe, which is very close to that of common basaltic rocks (for example, of diorite). However, since September 1967, when Surveyor 5 landed successfully in Mare Tranquillitatis, we learned for the first time the actual proportions of the principal elements constituting the lunar surface from the way in which α-particles are back-scattered by it. This experiment—repeated two months later by Surveyor 6 in Sinus Medii (Plate 154), and in January 1968 by Surveyor 7 in the proximity of the crater Tycho (Plate 199)—disclosed that the atomic composition of the lunar surface in all three places appears to be quite uniform (the only noticeable difference being a somewhat lower iron content in the continental area near Tycho), and a similarity was noted with the atomic composition of the terrestrial basalts.

The α-particle experiments aboard these three Surveyors could indicate nothing about the molecular structure of lunar matter (nor about the heavy elements present in low abundances). This information did not get in our hands until the return of the Apollo 11 mission in July 1969, with the 22 kg of lunar material to be analyzed in the laboratory. Only preliminary results of this analysis are available as this atlas goes to press; but their outcome makes us gasp not only at the achievement itself but even more at its implications.

The most abundant constituent of the material brought back by Apollo 11 from Mare Tranquillitatis is oxygen (40–60 per cent by weight); aluminum (4–7 per cent), titanium (4–6 per cent) and magnesium (4–6 per cent)—to list only elements present in amounts exceeding 1 per cent (the ranges indicated for each element refer to the respective abundances in different samples). The samples returned by Apollo 12 from the Oceanus Procellarum in November 1969 were found to be of a broadly similar composition with some differences (such as a lower titanium content).

The molecular composition of lunar matter was found to consist predominantly of silica (SiO_2; 38–43 per cent) followed by FeO (16–21 per cent), Al_2O_3 (9–13 per cent), CaO (9–12 per cent), TiO_2 (7–11 per cent),

MgO (7–10 per cent) and other constituents amounting to less than 1 per cent by weight. The bearing strength of the surface amounts to a few pounds per square inch (that is, about 3×10^4 dynes/cm²), increasing rapidly with depth; and the bulk density of the fine-grain component of the material is between 1.5–1.6 g/cm³. Since the densities of the individual compact grains were found to range between 3.1–3.5 g/cm³, it follows that about half of the volume of the bulk of material lifted from the topmost layer of the surface is empty space. This accounts for the relatively low bearing strength of the material, and for its low effective dielectric constant. Therefore, the bulk of the material covering the lunar surface as we see on the photographs reproduced on Plates 192–194, in which the footpads of the Surveyor sank down on landing (Plate 192), or in which the astronauts made imprints of their footsteps (Plate 214), consists of loosely packed silicate material (predominantly silica).

How do these results compare with expectations? That the surface of the Moon is covered with loosely packed material of very low thermal conductivity and radar reflectivity was known from ground-based observations on the Earth; its relatively low bearing strength had been previously disclosed by the Surveyors. However, prior to the return of the Apollo samples there were two schools of thought about the chemical composition of lunar material. One school expected the Moon to possess a chemical composition akin to that of the terrestrial mantle (which would have been the case if the Moon's mass had ever been detached from the Earth). The other—regarding the Moon as a primary object possibly older than our planet—anticipated a similarity between the composition of the Moon and that of the solar atmosphere, as the latter should approximate as closely as anything within observational (spectroscopic) reach the unadulterated composition of primordial matter from which the solar system originated (unadulterated because the large mass of the Sun prevented any selective escape of the elements from its gravitational field and because its temperature is too low to cause nuclear transformations on any appreciable scale).

The observational verdict delivered since 1967 by the lunar spacecraft confounded both these views, and presented the Moon to us as a much more enigmatic celestial body. Chemically, the Moon proved to be quite different from both the Sun and the Earth; for example, its content of titanium (and also chromium, zircon, and others) is very much higher than in the Sun's atmosphere or the Earth's crust; whereas other elements (like nickel and sodium, potassium or europium) are again very much less abundant. In particular, the ratio of iron to nickel in the Moon appears to be larger than that encountered in any other sample of cosmic matter we know (the Earth's crust; solar atmosphere; meteorites); and common elements like carbon or nitrogen appear to be conspicuously absent from the compounds found so far in the lunar crust—the former not being there in quantities exceeding one part in ten thousand by weight. Water seems likewise to be completely missing from all samples analyzed so far.

What kind of rocks have been found on the Moon? All those brought back so far are igneous, of generally basaltic composition; and numerous minerals of this type well-known from the Earth, such as olivine, plagioclase feldspar, ilmenite, and others, have been identified in many samples. In point of fact, only three new minerals, not known previously from the Earth, have been found in lunar rocks to date. Their crystalline structure and chemical properties indicated, moreover, that lunar rocks solidified at temperatures between 1,000–1,200°C under highly reducing conditions (the partial pressure of available free oxygen had to be less than 10^{-13} of an atmosphere to account for the virtual absence of higher states of oxidation). Moreover, many rocks exhibit evidence of shock metamorphism, which is strongly suggestive of effects of the passage of intense shock-waves through solids—such as can be produced on the Moon by meteoritic impacts from space.

All rocks returned from the Moon so far appear to be—we repeat—igneous;

but the reader should not jump from this to a conclusion equating "igneous" with "volcanic." All rocks we call "volcanic" on the Earth are, to be sure, igneous; but the converse is not necessarily true. All lunar rocks we possess indicate the effects of heat treatment; but certainly no rocks like these ever passed through the crater of any terrestrial volcano!

What could, then, be their origin, and what is the origin of the Moon as a whole? In the present state of research it appears probable that the Moon—like most other bodies of the solar system—came into being by an agglomeration of solid preexisting particles; for it does not seem possible to envisage a workable process which could lead to the formation of planetary bodies—let alone of mass as small as that of the Moon—by condensation of gas at moderate or high temperature. Therefore, the view prevalent today is that the Moon accumulated from preexisting solid particles more than 4,000 million years ago; and it is at least possible that some of these particles may have been heated up to temperatures well above their melting point by the Sun, which at that time may have been in the last stage of its contraction toward the Main Sequence, and very much larger than now—possibly of dimensions comparable with those of the present orbits of inner planets.

At this stage the Sun would have been somewhat cooler than today; but dust clouds exposed to it at a shorter range could have been heated to several hundred degrees or even more. If any particles evaporated during this heat treatment, the gas thus produced would have dissipated beyond retrieval; but particles which only melted could have cooled off again quite rapidly in a condensing swarm (in which the individual particles shield each other by their shadows from the scorching sunlight). It is, therefore, possible that not only the surface material, but the bulk of the Moon's mass may have acquired its igneous (basaltic) nature *before* the original swarm of solid particles coalesced into the lunar globe as we know it today; and that the source of heat responsible for this conversion need not have been any radioactively heated volanic "pockets" below the surface, but radiant energy of the youthful Sun.

Perhaps the most important and profound scientific result which we owe to the Apollo missions has been a determination of the *time* which has elapsed since the solidification of the lunar rocks now in our hands. As is well known, this age can be ascertained from continuous ticking of atomic clocks embedded in most minerals occurring in nature—in the form of radioactive elements spontaneously disintegrating at a known rate. The particular atomic clock best suited for measurements of long intervals of time is radioactive potassium (^{40}K), which disintegrates (by β-decay) into a kind of argon (^{40}Ar) at a constant rate. Laboratory measurements have disclosed this rate to be such that one-half of potassium 40 will decay into argon in a time close to 1,270 million years, which is quite independent of the physical conditions (temperature, pressure) to which solid minerals are exposed.

The decay product—argon—is a gas which would escape freely from a liquid, but becomes imprisoned in the crystal lattice of a mineral from the moment it solidifies. If, at a later time, we grind a sample to powder and expel gas from it by heating, the ratio of the decay product (^{40}Ar) to its mother substance (^{40}K) will indicate the time that has elapsed since the respective sample solidified as the hand of a cosmic clock and neither temperature nor pressure to which rocks can be exposed will alter in the least the regularity of its march.

Other chains of spontaneous radioactive disintegrations—such as uranium-lead or rubidium-strontium decays—can be used for the same purpose; but they all possess one feature in common: namely, they all start marking the time from the moment when the respective rock solidified; their dials would be automatically set back to zero on remelting. And, we may add, when applied to the lunar rocks they all told essentially the same story.

When such methods were applied recently to the determination of the age of crystalline lunar rocks which the Apollo 11 astronauts brought back to the

Earth from Mare Tranquillitatis, they indicated that *these rocks must have crystallized 3,000 to 4,000 million years ago*—with the average age close to 3,700 million years. On the other hand, similar finds picked up by Apollo 12 at its landing place in Oceanus Procellarum proved to be about 300 million years younger. The dispersion in age within each group is no doubt real; and testifies to the fact that not all rocks found in the same place on the Moon actually solidified *in situ*. The local "fines" represent rather a mixture of rocks which may have been transported there (by impact throwouts) from different parts of the Moon. A difference between the average age of the majority of rocks brought back from Mare Tranquillitatis and Oceanus Procellarum is probably due to a difference of about 300 million years between the events which gave rise to these formations—Mare Tranquillitatis being the older of the two.

Perhaps the most interesting result that has come from radioactive dating of lunar rocks is the fact that, on each landing site investigated by Apollos 11 and 12 so far, a quantity of smaller debris was found to exhibit substantially greater age. The time which elapsed since the solidification of these small "fines" proved, in fact, to cluster around 4.6 billion years in *both* localities—making their ages virtually identical with those of the oldest known meteoritic material. A lighter color (that is, higher albedo) of many of these small chips lends some weight to a conjecture that these chips solidified in the continental regions of the lunar surface, and were subsequently transported to their present localities by the mechanical action of meteoritic impacts.

Whatever may have happened, however, these chips constitute irrefutable testimony that *solid matter already existed on the lunar surface 4.6 billion years ago,* which has not since been melted, while the substantially younger age of dark rocks characteristic of the lunar surface from the plains of Mare Tranquillitatis or Oceanus Procellarum discloses that later solidification of material in these localities must have been the result of subsequent but isolated events. This is indeed what we inferred earlier from the stratigraphy of the lunar face; and the radioactive dating of the rocks collected at two localities so far has provided our time-scale with absolute calibration. Earlier in this text we surmised the great age of the stony relief of the lunar surface from the number of its disfiguring pockmarks. Now we know that the continental land masses may have solidified while the Sun was in the last throes of its Kelvin contraction toward the Main Sequence.

Nothing of comparable age can certainly be found anywhere on the Earth, or (as far as we know) elsewhere in the solar system. In contrast with the Moon, our Earth exhibits to the outside world a cosmic face of almost eternal youth—rejuvenated continuously by geological processes such as erosion and denudation of its land by the combined action of air and water and (more important) by continuous continental drift operative in its mantle and driven by the internal heat engine of the Earth. Very few parts of exposed terrestrial continents, or even ocean floors, are now known to be older than a few hundred million years. In contrast, the Moon (on account of its small mass and heat capacity) can afford none of these means of cosmic cosmetics to make up her face. Therefore the latter mirrors truly the ages gone by and preserves a reflection of events that occurred long before our own terrestrial continents were formed, and long before the first manifestation of life on Earth flickered in shallow waters. As a monument from the past, the Moon constitutes the most important fossil of the solar system; and a correct interpretation of the hieroglyphs engraved on its stony face holds indeed a rich scientific prize.

A search for the presence of unstable nuclides in the lunar crust produced by cosmic rays led, moreover, to an interesting by-product in the form of a determination of the rate at which the lunar surface is being "plowed over" by meteoritic infall. Microchemical analysis of samples of lunar rocks from Mare Tranquillitatis brought back by Apollo 11 disclosed the presence in them, not only of many stable elements present elsewhere in nature, but also

4. Surface of the Moon and Its Structure

of certain unstable elements (like ^{26}Al), which are known to be produced by impact of hard corpuscular radiation commonly referred to as cosmic rays. These rays are of partly solar, partly galactic origin; and their flux remains roughly constant in time.

The penetrating power of cosmic rays but seldom exceeds about one-half to one meter of the lunar crust. If, therefore, we find rocks on the Moon showing signs of exposure to cosmic rays (such as the presence in them of nuclear spallation products which do not otherwise occur in nature) at some depth underground, it follows that these rocks must once have been exposed to cosmic rays closer to the surface. Moreover, from the state of decay of these spallation products we can infer the length of time which elapsed since the gradual underground burial of such rocks.

A determination of such "cosmic-ray exposure ages" of lunar rocks brought back by Apollo 11 has shown that *these rocks must have been within one meter of the surface for periods ranging from 20 to 160 million years—* with an average of close to 100 million years. Cumulative effects of mechanical disturbances associated with cosmic impacts on the Moon over longer intervals of time may bury rocks underground to subsurface depths to which cosmic rays can no longer penetrate in appreciable flux; but a topmost layer about one meter in depth seems to have been plowed over by cosmic erosion in a time-interval of the order of 100 million years, that is, about the time separating us from the Cretaceous geological period on the Earth.

It is indeed a sobering thought to consider that the footprints, which the Apollo astronauts and their followers will have left in the soft lunar ground, may possess a degree of permanence comparable to the footprints of the dinosaurs in the soft ground of the Cretaceous period, preserved for us by fossilization. Should, perchance, human civilization perish tomorrow by some disaster and journeys to the Moon not be resumed for millions of years, the chances are good that our distant descendants of that age would find footprints there still much the same as we see them now on Plate 214 of this atlas. These footprints (and other marks of human activity) on the Moon may, therefore, yet prove to be a better claim to immortality for the intrepid travelers who made them than all their fleeting glory on the Earth.

II. ATLAS OF THE MOON WITH DESCRIPTIONS OF THE PLATES

The principal part of this atlas consists of 214 plates illustrating all essential features of the lunar surface, the individual contents of which will be described below.

As the reader will readily notice, the entire content of this work divides more or less into six distinct groups. The first one (Plates 1–12) features the "tools of research" from ground-based telescopes used by Bernard Lyot in the 1940's to several generations of lunar spacecraft ending with Apollo 12 on the Moon in November 1969. Plates 13–75 will take us on a guided tour over the principal types of lunar ground—mainly maria—on the visible hemisphere of the Moon, with occasional excursions to the Moon's far side, as seen from the distance of the Earth as well as from closer proximity of the lunar surface taken along a north-west-south-east circuit. On Plates 76–154 we shall repeat the same tour (starting from south toward the east) to inspect the principal types of continental ground with craters encountered on the Moon; and Plates 155–181 will similarly illustrate the principal examples of formations known as clefts, domes and rilles.

All the lunar photographs reproduced on Plates 16–181 feature landscape at the time of sunrise or sunset on the Moon, for the shadows cast in the rays of the rising or setting Sun bring out the vertical relief of the landscape. When the Sun stands high above the horizon, the lunar landscape assumes, however, a rather different aspect—as illustrated on several examples on Plates 182–189. The last group of plates—190–214—then exhibit various features of more detailed structure of the lunar surface on a meter, centimeter, and millimeter scale, ending up with imprints left by the first men on the Moon in July 1969.

In contrast to all atlases of the Moon which have appeared in the past, and in which views of its landscape were invariably reproduced as seen through astronomical inverting telescopes, all the ground-based photographs of the Moon included in this atlas are reproduced so that north is (approximately) on the top and east to the right—except for a few photographs taken from orbiting spacecraft, where the position of the horizon depended on the spacecraft's immediate altitude. In other words, the orientation of the photographs reproduced in this atlas corresponds essentially to a naked-eye view of the Moon from the Earth, or from a spacecraft approaching it.

A brief description of each photograph reproduced in this atlas is given in the following remarks. The top of each plate is indicated by an arrow, shown here:

1. Observatoire du Pic-du-Midi in the French Pyrénées (2,861 m above sea level), where many of the photographs reproduced in this atlas have been made since 1964 as part of a collaborative program between the Department of Astronomy, University of Manchester, England and the U.S. Air Force.

2. The observatory's 24-inch refractor, of 18.22 m focal length, used extensively for lunar photography at Observatoire du Pic-du-Midi since the days of Bernard Lyot.

3. The 43-inch Manchester-Observatoire du Pic-du-Midi telescope of 16.05 m focal length (f/15), with which the majority of the ground-based photographs of the Moon reproduced in this atlas were taken between 1964 and 1966.

4. The U.S. Ranger spacecraft (with its solar panels folded in position for launch), in the workshops of the Jet Propulsion Laboratory, California Institute of Technology, where it was designed and built. The cylinder on top is the omnidirectional antenna; the inset (upper left) shows the location of its optics. (Reproduced by courtesy of the Jet Propulsion Laboratory, California Institute of Technology)

5. The U.S. Surveyor spacecraft in the laboratory—a prototype configuration of five spacecraft which succeeded in making soft landings on the Moon. The periscope device which served as the spacecraft's eyes, and televised to the Earth many of the views reproduced in this atlas, can be seen on the left of the main mast carrying the solar panels. (Reproduced by courtesy of NASA and of JPL)

6. Surveyor on the Moon. The outlines of a surface sampler aboard Surveyor 3 appear on a photograph televised by this spacecraft on May 2, 1967 together with the shadow cast by it on the lunar surface by the rays of the late afternoon sun. The arm of the sampler can reach up to a distance of 1.5 m from the spacecraft. In between the sampler and the main mast, can be seen the shadow of the TV camera which took this picture. (Reproduced by courtesy of NASA and JPL)

7. A model of the Russian spacecraft Luna 9 which effected the first soft land on the lunar surface on February 4, 1966. The numbers of individual parts marked on the plate refer to: 1) soft-landing portion of the unit (for a fuller view of it, see Plate 8); 2) guidance system compartment; 3) camera system; 4) electronic portion of altitude control system; 5) liquid-fuel retrorocket engine; 6) altitude control jets; 7) spherical fuel tank; 8) oxidizer; 9) vernier jet engines; 10) gas supply for altitude control jets; 11) radio-altimeter; and 12) directional parabolic antenna. (Reproduced by courtesy of the USSR Academy of Sciences)

8. A model of the 100-kg capsule of the Russian spacecraft Luna 9 with antenna and optics unfolded. The sperical part of the spacecraft is approximately half a meter in size. (Reproduced by courtesy of the USSR Academy of Sciences)

9. A view of the U.S. Orbiter spacecraft in the laboratories of the Aerospace Division of the Boeing Company, with antennae and solar panels unfolded. The optics of the Orbiter's cameras are exposed to view in front. (Reproduced by courtesy of NASA and of the Boeing Company, Seattle, Washington)

10. Men on the Moon. Edwin Aldrin leaving the Lunar Excursion Module "Eagle" on July 20, 1969 to descend onto the lunar surface. (Reproduced by courtesy of NASA)

[66]

PLATE 1

PLATE 2

PLATE 3

PLATE 4

PLATE 5

PLATE 6

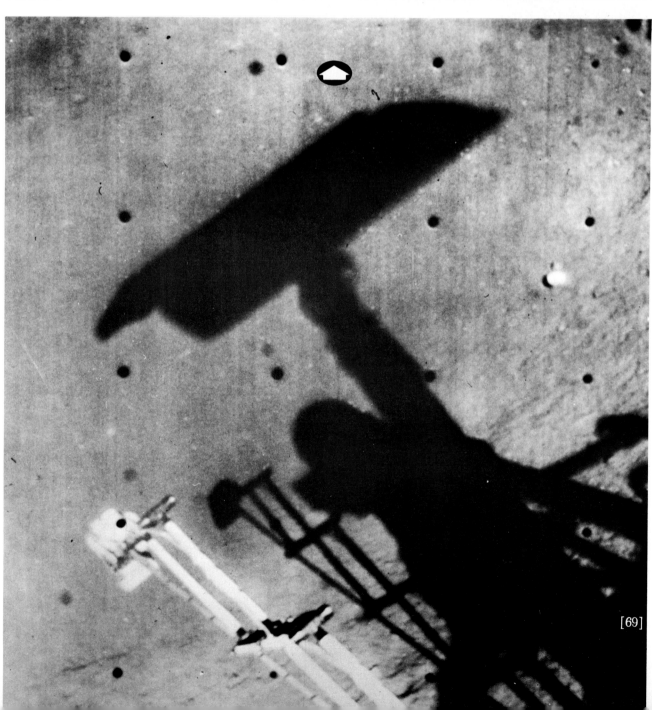

[69]

PLATE 7

PLATE 8
[70]

PLATE 9

PLATE 10

11. Tranquillity Base on July 20, 1969. The "Eagle" spacecraft behind the seismometer (in front). (Reproduced by courtesy of NASA)

12. Rendezvous of men and spacecraft on the Moon on November 20, 1969: Astronaut Charles Conrad standing beside Surveyor 3 (which landed in Oceanus Procellarum, near the crater Lansberg, on April 20, 1967); with the Apollo 12 Excursion Module "Intrepid" in the distance within 200 m. (Reproduced by courtesy of NASA)

13. A view of the full Moon as seen from Earth and photographed with the 24-inch refractor at Observatoire du Pic-du-Midi on February 18, 1962. The photograph shows clearly the distribution of the lunar continents and the maria on the hemsphere facing the Earth. (Manchester Lunar Program)

14. A view of the Moon after first quarter, photographed with the 24-inch refractor at Observatoire du Pic-du-Midi on March 30, 1966. (Manchester Lunar Program)

15. A view of the Moon at a time close to the last quarter, as photographed with the 24-inch refractor at the Observatoire du Pic-du-Midi on 6 October 1966. (Manchester Lunar Program)

16. Sunset over Mare Serenitatis and the eastern shores of Mare Imbrium with the craters Aristillus, Archimedes, and Autolycus; Mare Frigoris north of the Alps; and the craters Aristoteles and Eudoxus north of Mare Serenitatis. Photograph taken on August 18, 1965 with the 24-inch refractor at the Observatoire du Pic-du-Midi. (Manchester Lunar Program)

17. Sunrise over Mare Serenitatis. Palus Putredinis of Mare Imbrium (with the craters Archimedes, Aristarchus, and Autolycus) in the upper left-hand corner; the craters Manilius, Menelaus, and Bessel at the bottom. Photograph taken on July 25, 1966 with the 43-inch reflector at the Observatoire du Pic-du-Midi. (Manchester Lunar Program)

18. Sunrise over the lunar Alps and Palus Nebularum of Mare Imbrium, with the craters Aristillus (bottom) and Cassini. The western part of Mare Frigoris (with craters Protagoras and Archytas) is on top. The first rays of sunrise begin to illuminate the rims of the crater Plato (top left). Photograph taken on September 22, 1966 with the 43-inch reflector at Observatoire du Pic-du-Midi. (Manchester Lunar Program)

19. Sunrise over Palus Putredinis of Mare Imbrium, dominated by the craters Archimedes, Aristillus, and Autolycus (center of field) and Cassini (upper right). Photograph taken on September 22, 1966 with the 43-inch reflector at the Observatoire du Pic-du-Midi. (Manchester Lunar Program)

20. Early evening over the vast plains of Mare Imbrium west of Palus Putredinis, dominated by the craters Archimedes (right) and Timocharis (left). Photograph taken on September 17, 1965 with the 43-inch reflector at the Observatoire du Pic-du-Midi. (Manchester Lunar Program)

21. Sunrise over the western parts of Mare Imbrium with Sinus Iridum, Jura Mountains, and Sinus Roris beyond them with Rümker (extreme left). Photograph taken on January 4, 1966 with the 43-inch reflector at the Observatoire du Pic-du-Midi. (Manchester Lunar Program)

22. Sunset over the western part of Mare Imbrium and Sinus Iridum (above); photographed on September 20, 1965 with the 43-inch reflector at Observatoire du Pic-du-Midi. (Manchester Lunar Program)

PLATE 11

PLATE 12

PLATE 13

[74]

PLATE 14

PLATE 15

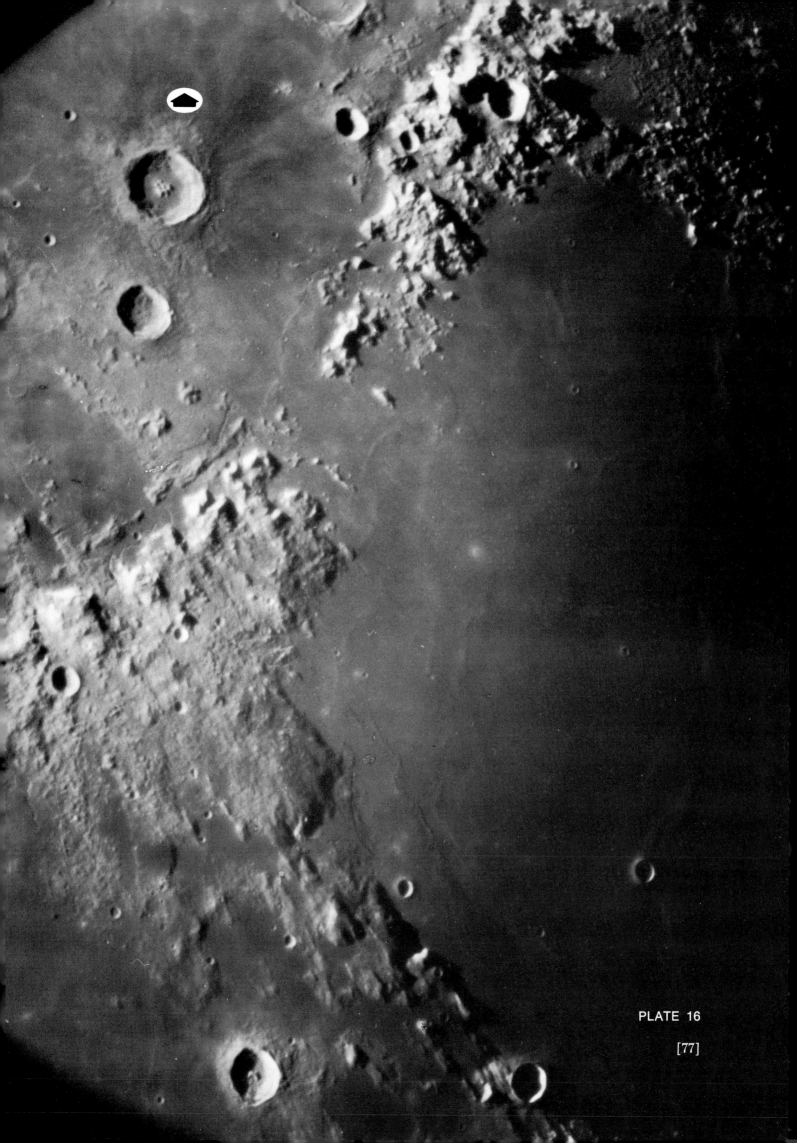

PLATE 16

[77]

PLATE 17

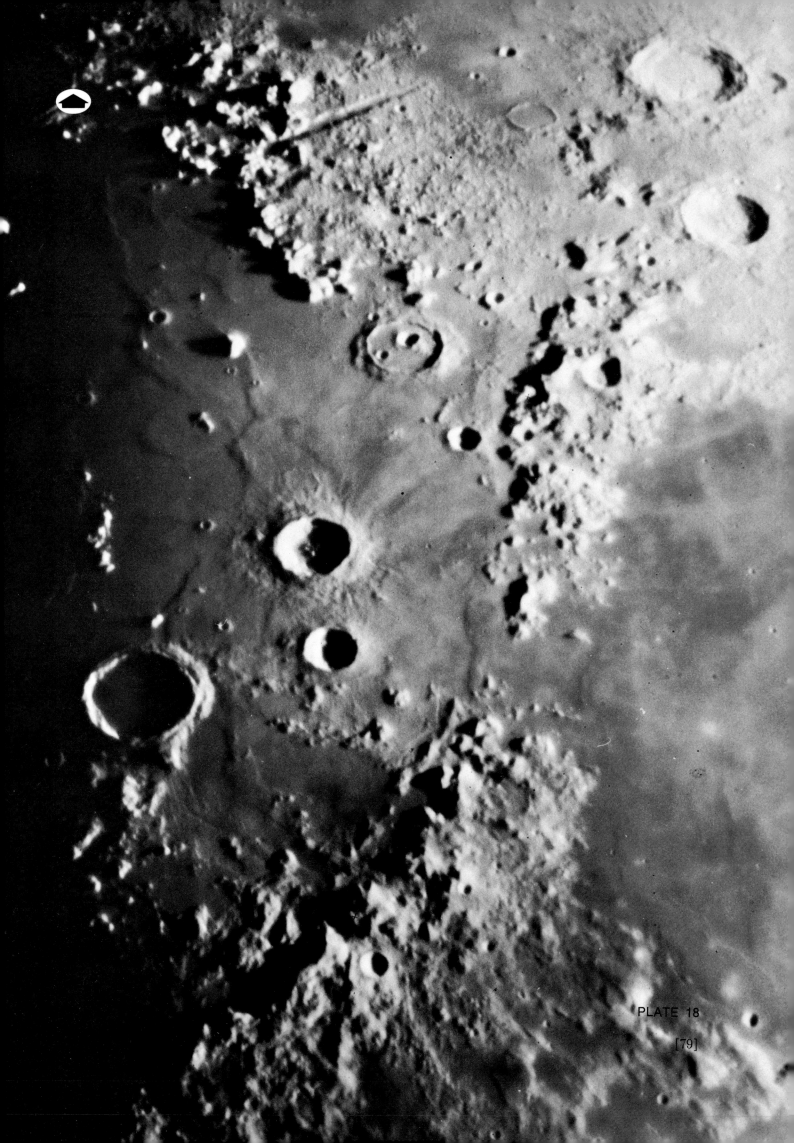

PLATE 18

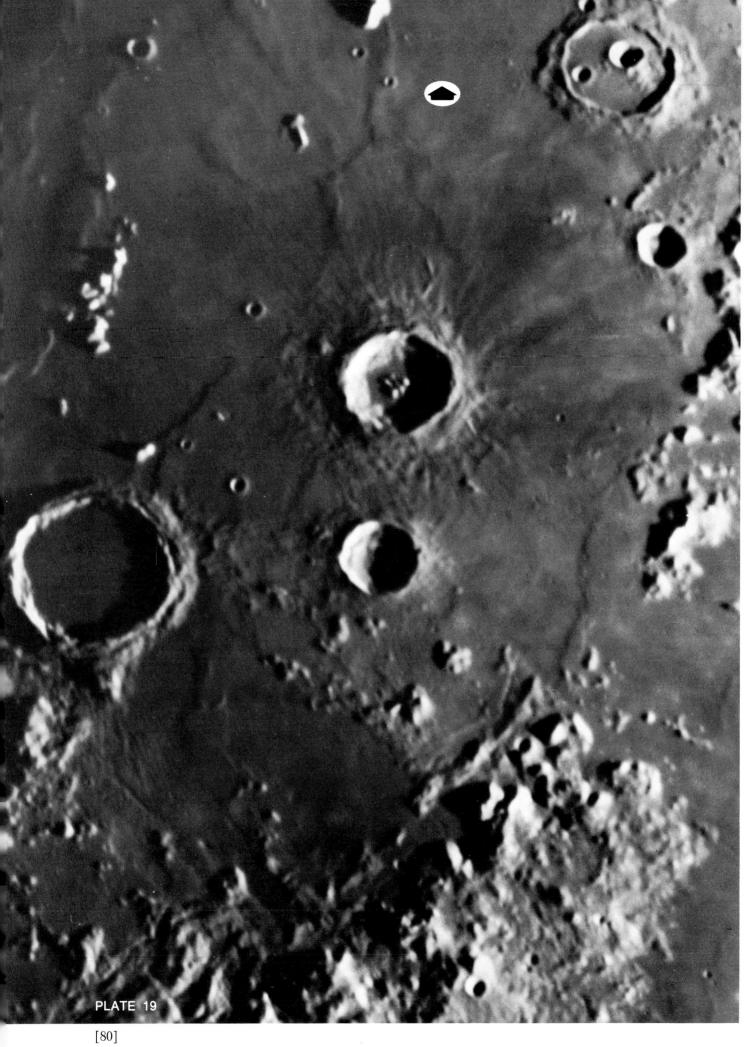

PLATE 19

PLATE 20

PLATE 21

PLATE 22

23. Sunset over the lunar Sinus Iridum, Jura Mountains and Sinus Roris, photographed on September 20, 1965 with the 43-inch reflector at Observatoire du Pic-du-Midi. (Manchester Lunar Program)

24. Sunrise over the western part of Mare Imbrium, Jura Mountains (top), and the Harbinger Mountains (bottom), dominated by the crater Mairan (top) and Prinz (bottom). Photograph taken on May 7, 1960 with the 24-inch refractor at Observatoire du Pic-du-Midi. (Manchester Lunar Program)

25. Western part of Mare Imbrium, with the craters Gruithuisen, Delisle (center), Diophantus and Euler (lower right). Photograph taken on May 7, 1960 with the 24-inch refractor at the Observatoire du Pic-du-Midi. (Manchester Lunar Program)

26. Western plains of Oceanus Procellarum near the crater Marius (center), Reiner and Seleucus are upper right; a group of distinctive low hills can be seen west of Marius. Photograph taken on April 3, 1966 with the 43-inch reflector at Observatoire du Pic-du-Midi. (Manchester Lunar Program)

27. Marius Hills—a group of small domes visible on the ground-based photograph reproduced on the preceding plate—as photographed by Orbiter 4 spacecraft on May 22, 1967 from an altitude of 2,667 km above the lunar surface. (Reproduced by courtesy of NASA-Langley and of the Boeing Company)

28. An enlarged section of the central part of the terrain shown on the preceding plate, showing domes, ridges and sinuous rilles, photographed by Orbiter 5 on August 18, 1967 from an altitude of 109 km. The width of the individual framelets corresponds to 3.5 km on the lunar surface; and white rectangles delimit areas shown at higher resolution on Plates 30–37. (Reproduced by courtesy of NASA-Langley and of the Boeing Company)

29. The region south of the one reproduced on the preceding plate, taken by Orbiter 5 on the same day, and with the same optics 15.2 seconds before Plate 28. (Reproduced by courtesy of NASA-Langley and of the Boeing Company)

30. High-resolution photograph of the head of a sinuous rille shown on Plates 28 and 29, taken by Orbiter 5 on August 18, 1967 from an altitude of 109 km above the lunar surface. The width of the individual framelets is approximately 400 m on the Moon. (Reproduced by courtesy of NASA-Langley and of the Boeing Company)

31. A high-resolution photograph of the arrowlike head of a rille shown on Plates 27 and 28 above the sinuous rille in the Marius Hills, taken by Orbiter 5 on August 18, 1967 from an altitude of 110 km. (Reproduced by courtesy of NASA-Langley and of the Boeing Company)

32. High-resolution photograph of the tail of the arrowheaded rille shown on Plate 29, taken by Orbiter 5 on August 18, 1967 from an altitude of 110 km. (Reproduced by courtesy of NASA-Langley and of the Boeing Company)

33. Details of the lunar surface to the north of the tail of the rille shown on Plate 32, photographed by Orbiter 5 on August 18, 1967 from an altitude of 111 km. (Reproduced by courtesy of NASA-Langley and of the Boeing Company)

PLATE 23

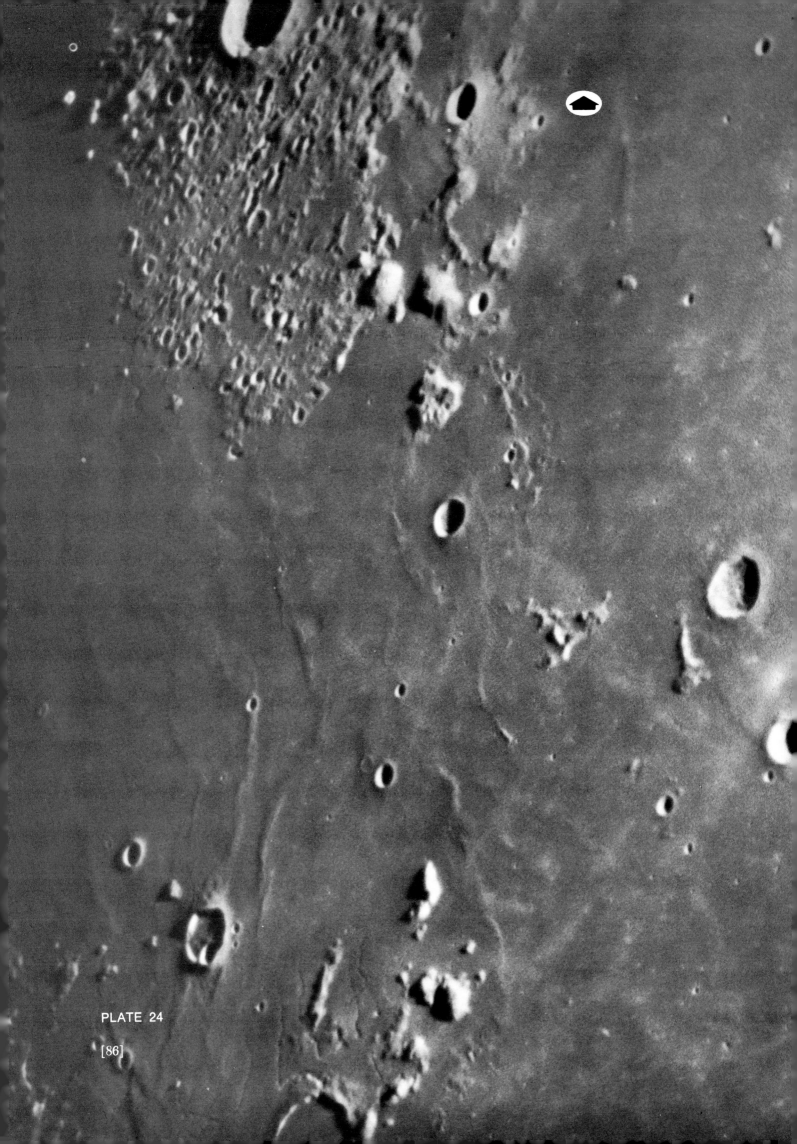

PLATE 24

[86]

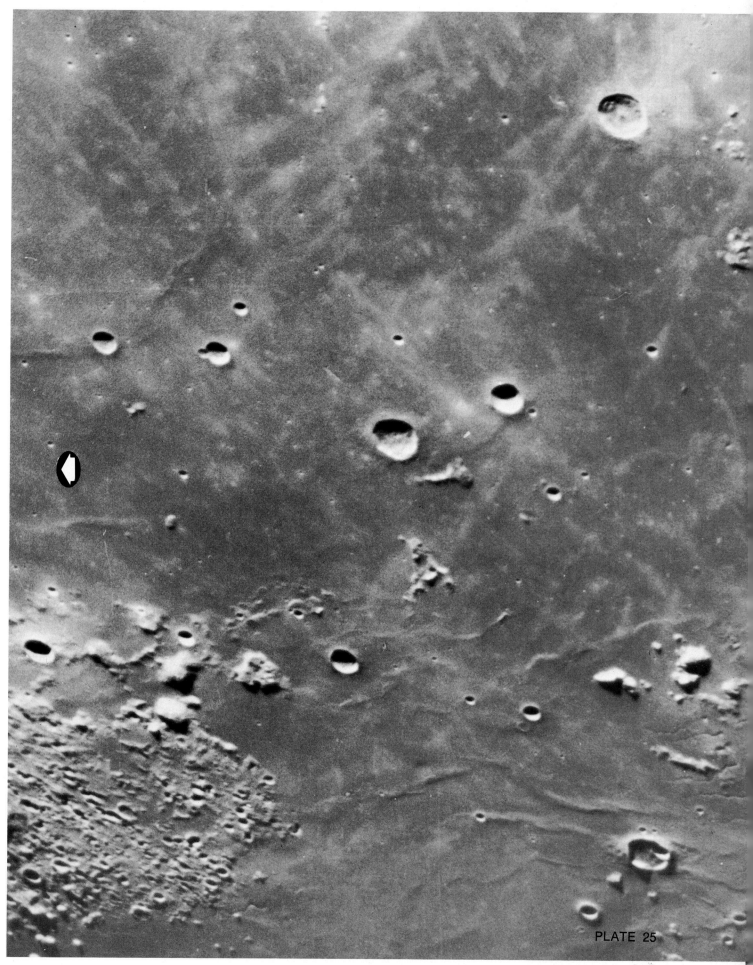

PLATE 25

[87]

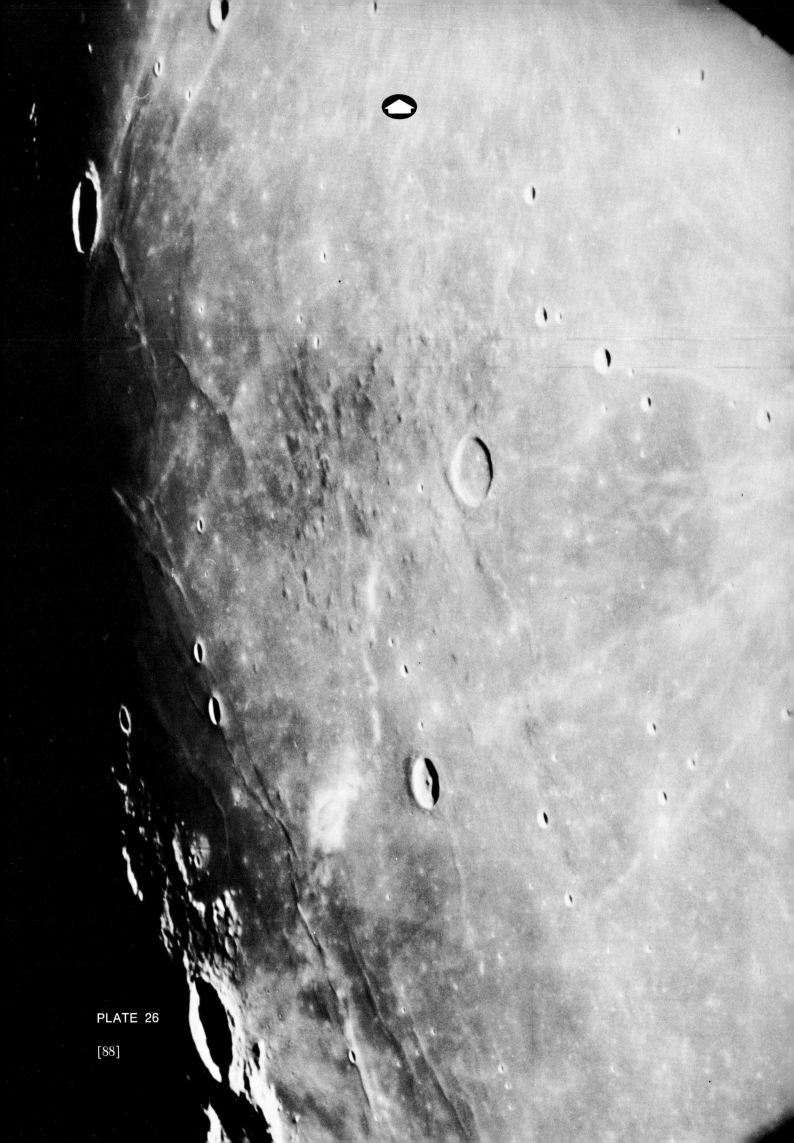

PLATE 26

[88]

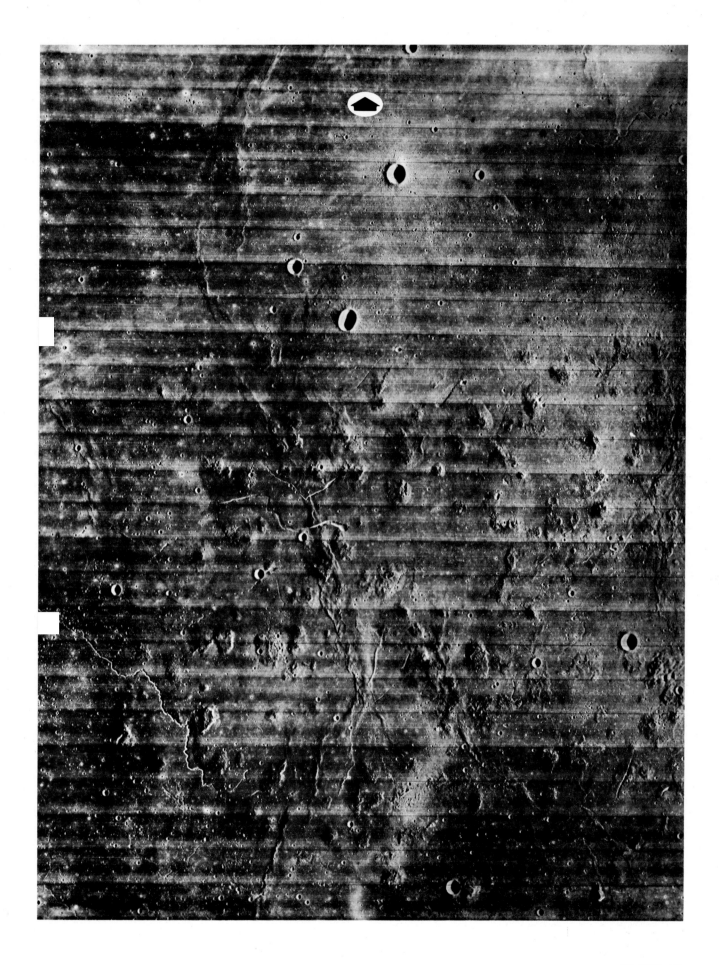

PLATE 27

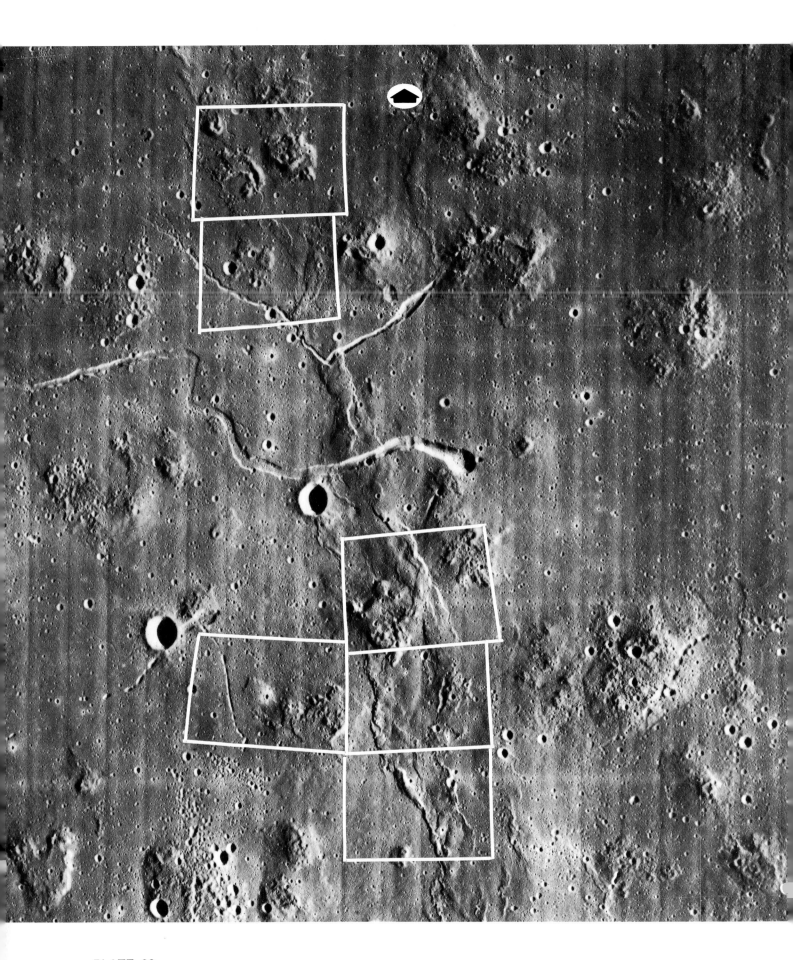

PLATE 28

[90]

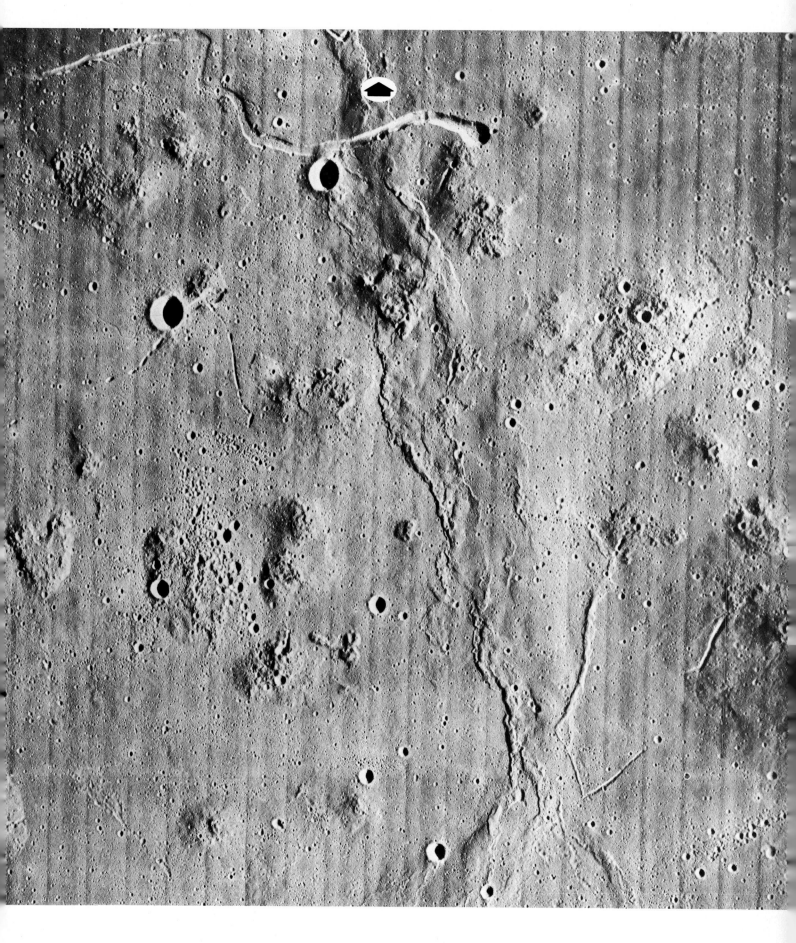

PLATE 29

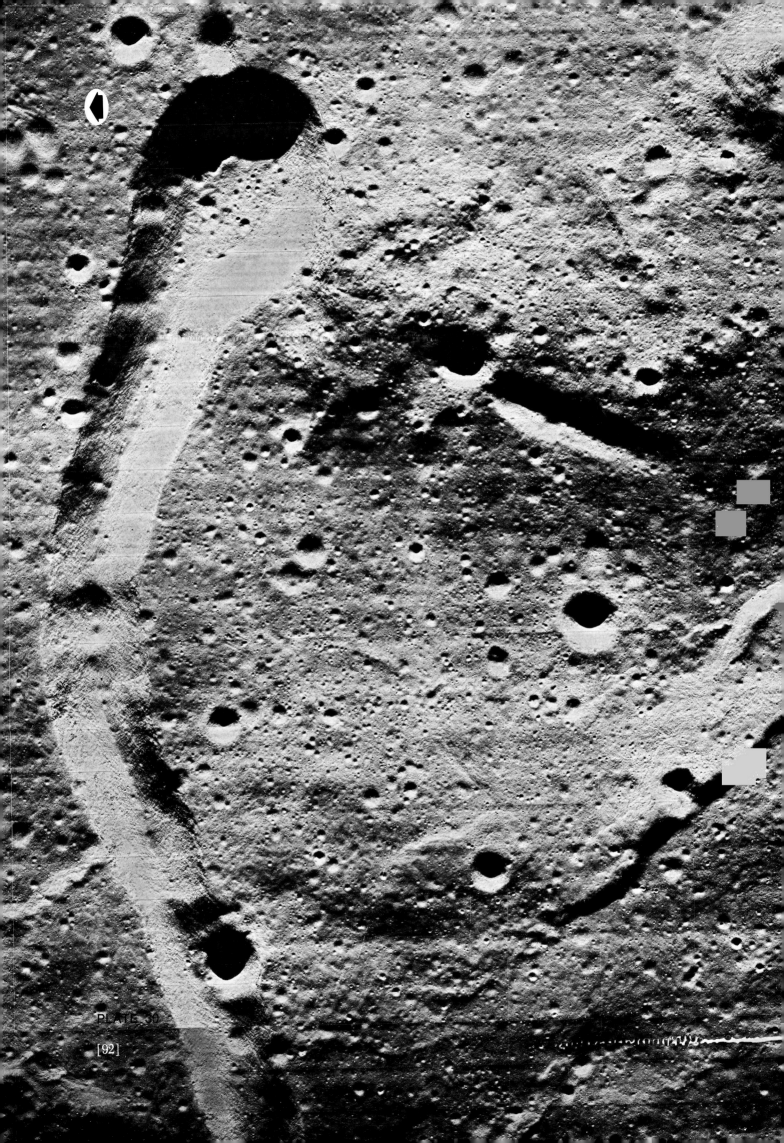

PLATE 50

[92]

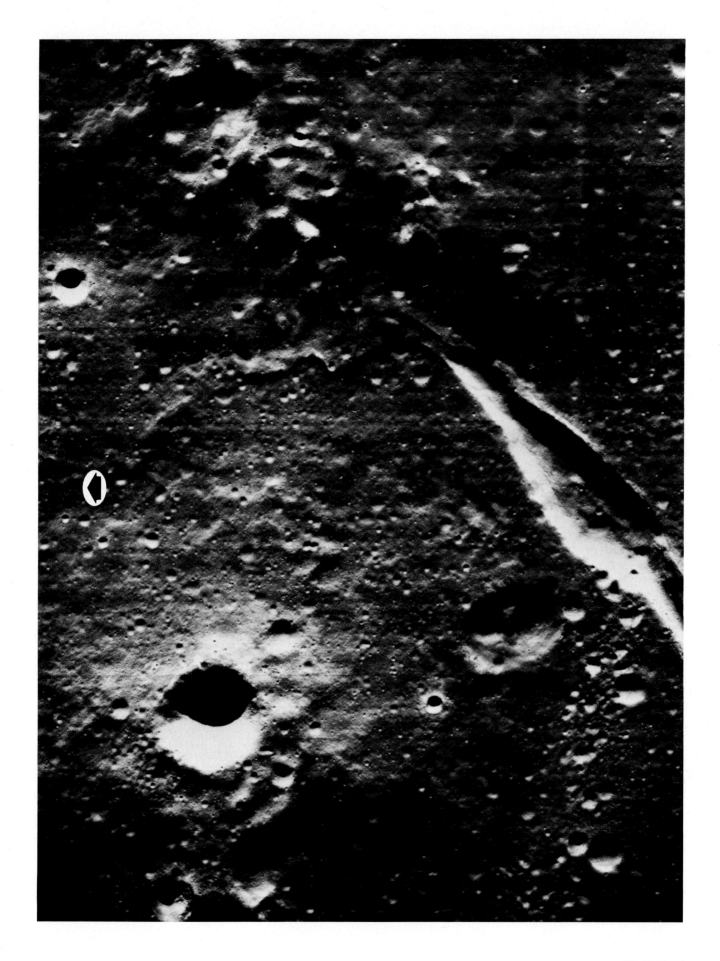

PLATE 31

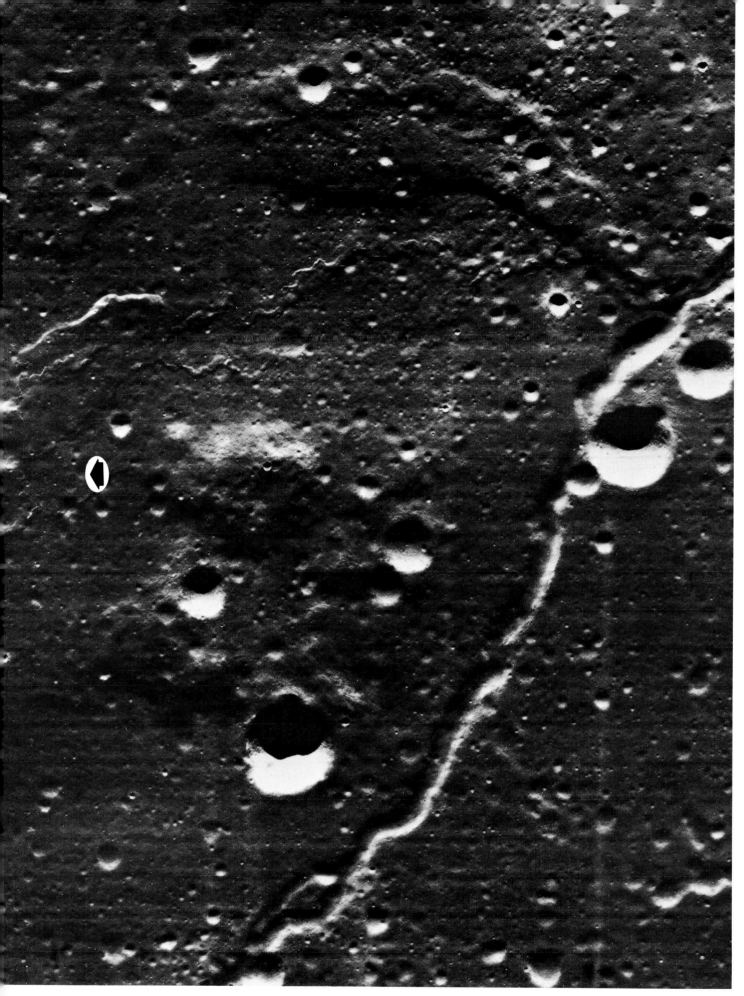

PLATE 32

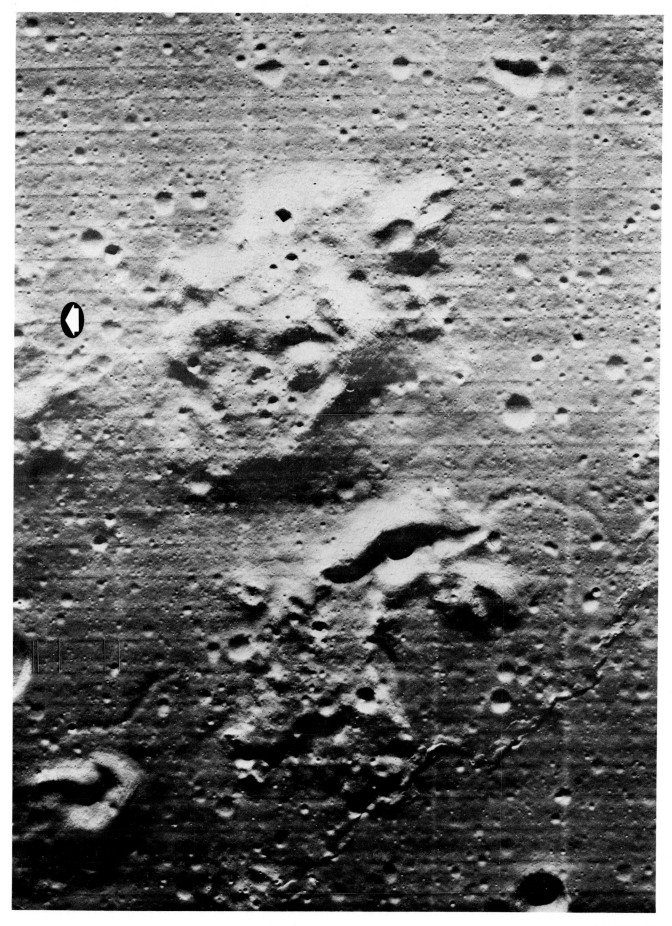

PLATE 33

34. Detail of the wrinkle ridge immediately to the south of the sinuous rille head seen on Plates 28 and 29 (just south of the high-resolution field of Plate 30), photographed by Orbiter 5 on August 18, 1967 from an altitude of 109 km. (Reproduced by courtesy of NASA-Langley and of the Boeing Company)

35. Detail of the surface to the south of the field shown on Plate 34, photographed by Orbiter 5 on August 18, 1967 from an altitude of 109 km. (Reproduced by courtesy of NASA-Langley and of the Boeing Company)

36. Wrinkle ridge to the south of the one shown on Plate 35; high-resolution photograph taken by Orbiter 5 on August 18, 1967 from an altitude of 109 km. (Reproduced by courtesy of NASA-Langley and of the Boeing Company)

37. Detailed photograph of a straight trench southeast of the sinuous rille shown on Plate 28 (and due east of the landscape shown on Plate 35), taken by Orbiter 5 on August 18, 1967 from an altitude of 108 km. (Reproduced by courtesy of NASA-Langley and of the Boeing Company)

38. An oblique view of the northern horizontal panorama of the region of the Marius Hills photographed by Orbiter 2 on November 25, 1966 from an altitude of 51 km above the lunar surface. (Reproduced by courtesy of NASA-Langley and of the Boeing Company)

39. Northern horizontal panorama of the vast plains of the Oceanus Procellarum, with the crater Marius (see also Plate 26) in the foreground. Photograph taken by Orbiter 2 on November 25, 1966 from an altitude of 48 km. (Reproduced by courtesy of NASA-Langley and of the Boeing Company)

40. Horizontal panorama of the northeastern plains of Oceanus Procellarum (east of the Marius Hills), with the Cavalerius Hills in the foreground, and three craters in the background, the largest of which is Galilei. Photograph taken by Orbiter 3 on February 22, 1967 from an altitude of 61 km. (Reproduced by courtesy of NASA-Langley and of the Boeing Company)

41. Cavalerius Hills (east of the crater Reiner, seen on Plate 26) as photographed by Orbiter 3 on February 22, 1967 from an altitude of 61 km. (Reproduced by courtesy of NASA-Langley and of the Boeing Company)

42. Details of the Cavalerius Hills in Oceanus Procellarum (see Plate 41), photographed by Orbiter 3 on February 22, 1967 from an altitude of 61 km. (Reproduced by courtesy of NASA-Langley and of the Boeing Company)

43. Cavalerius Hills from a different perspective: the northwest part of Oceanus Procellarum bordered by the craters Aristarchus-Herodotus (upper right corner), Kepler (bottom right), and the chain of Grimaldi, Lohrmann, Hevelius, and Cavalerius in the west (right). Photograph taken on February 3, 1966 with the 74-inch reflector of Helwan Observatory at Kottamia, Egypt. (Manchester Lunar Program)

44. Western parts of the Oceanus Procellarum (with Kepler and Aristarchus) at top right, and of Mare Humorum (bottom right) from an unusual perspective never visible to the human eye from Earth—as photographed with the wide-angle lens of Orbiter 4 on May 23, 1967 from an altitude of 2,720 km above the lunar surface. The craters Grimaldi, Lohrmann, Hevelius, and Cavalerius, distorted by terrestrial foreshortening on Plate 43, are seen from an almost overhead vantage point above the center of the plate. The large dark-floor crater seen near the bottom of the field is Schickard. The first rays of the rising Sun are beginning to illuminate the chain of the Cordillera Mountains—eastern ramparts of Mare Orientale. (Reproduced by courtesy of NASA-Langley and of the Boeing Company)

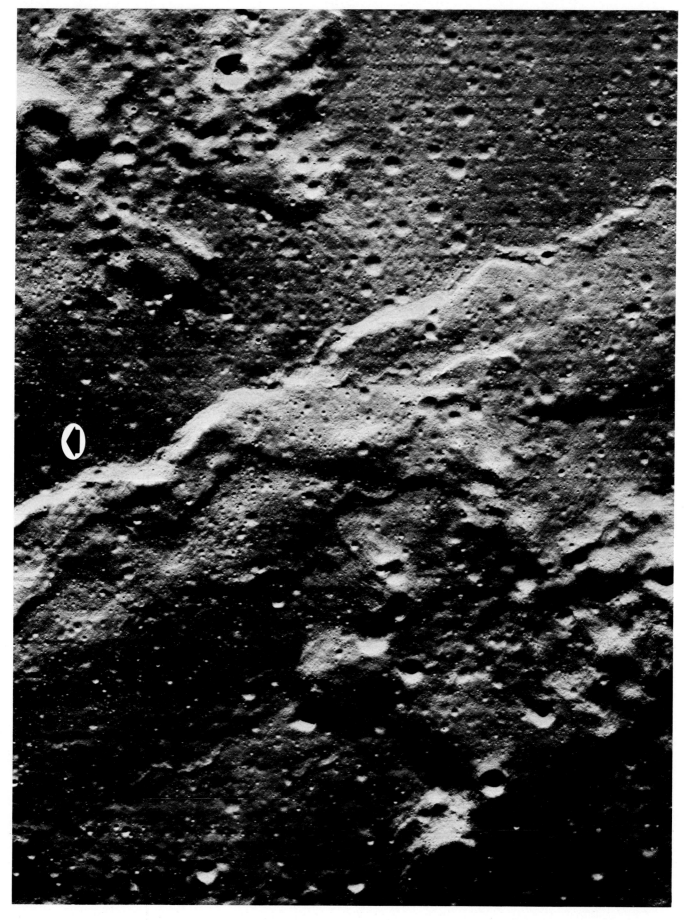

PLATE 34

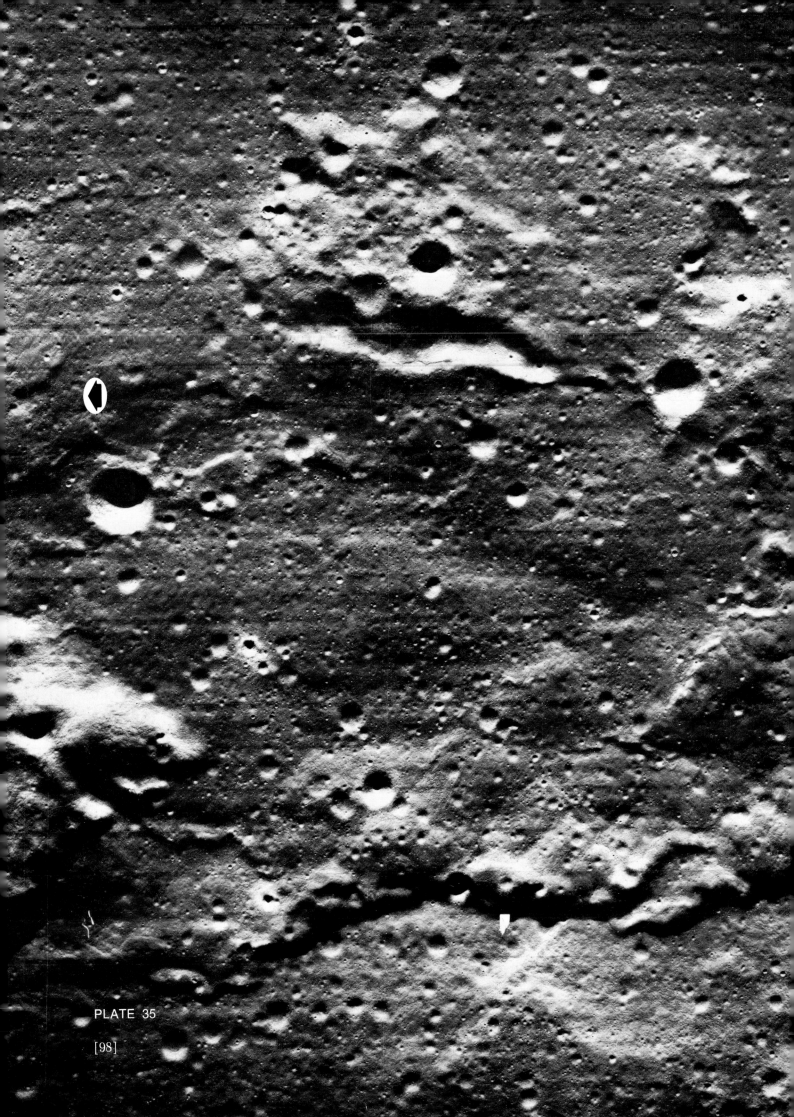

PLATE 35

[98]

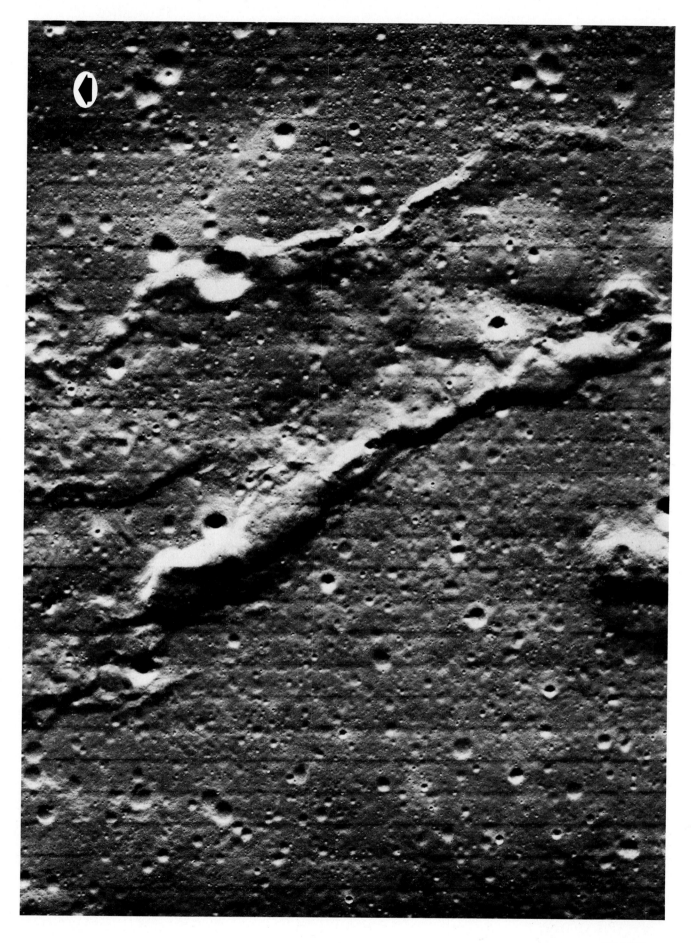

PLATE 36

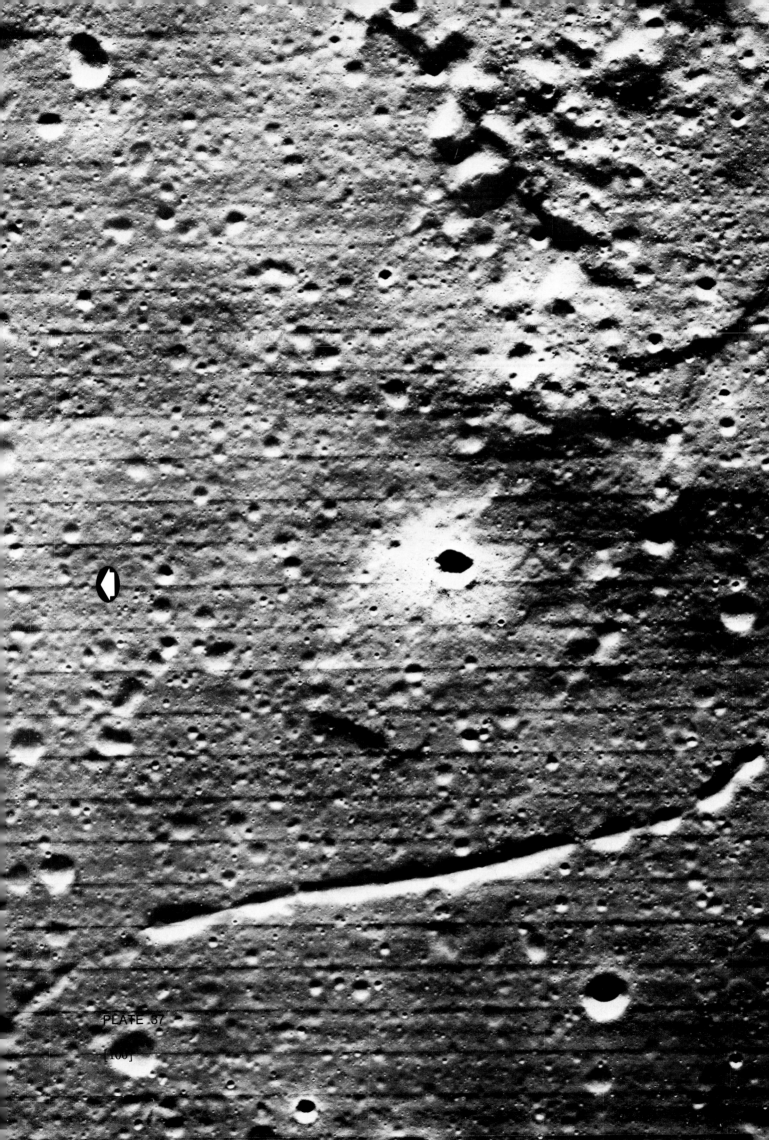

PLATE 37

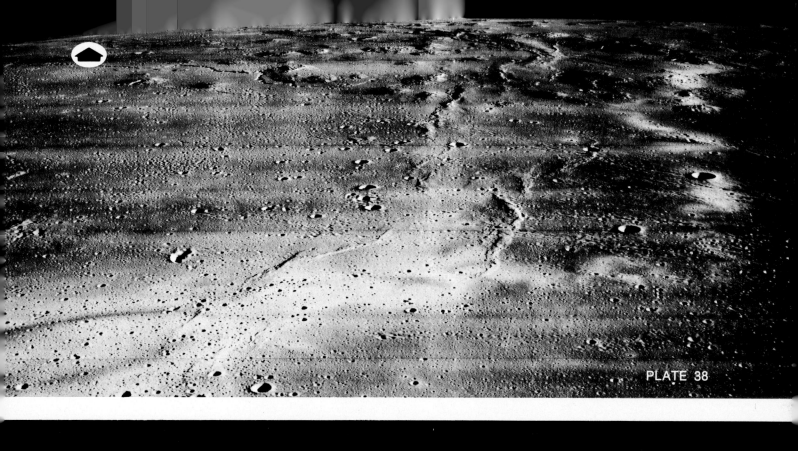

PLATE 38

PLATE 39

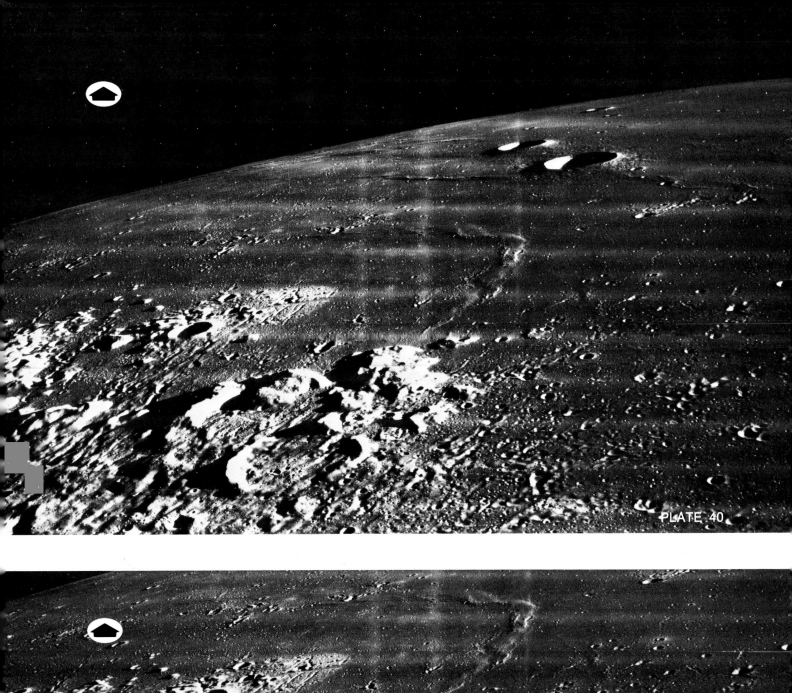

PLATE 40

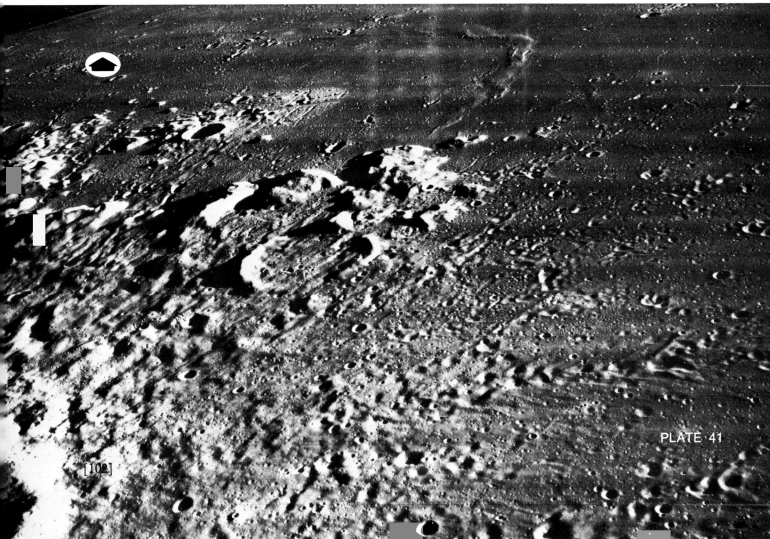

[102]

PLATE 41

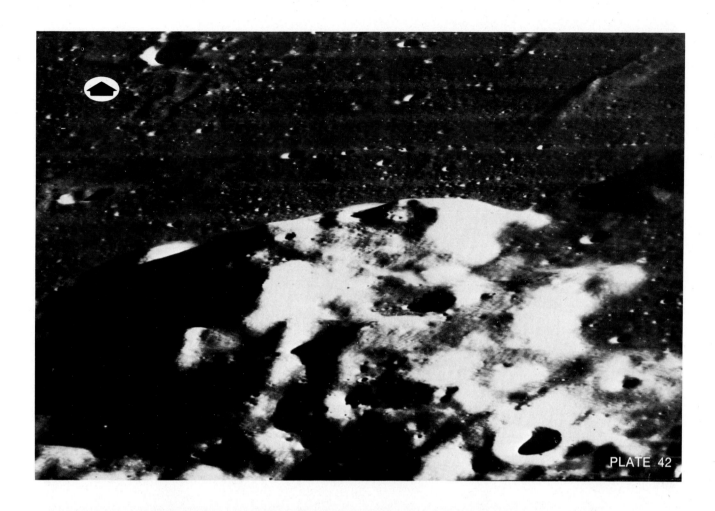

PLATE 42

PLATE 43

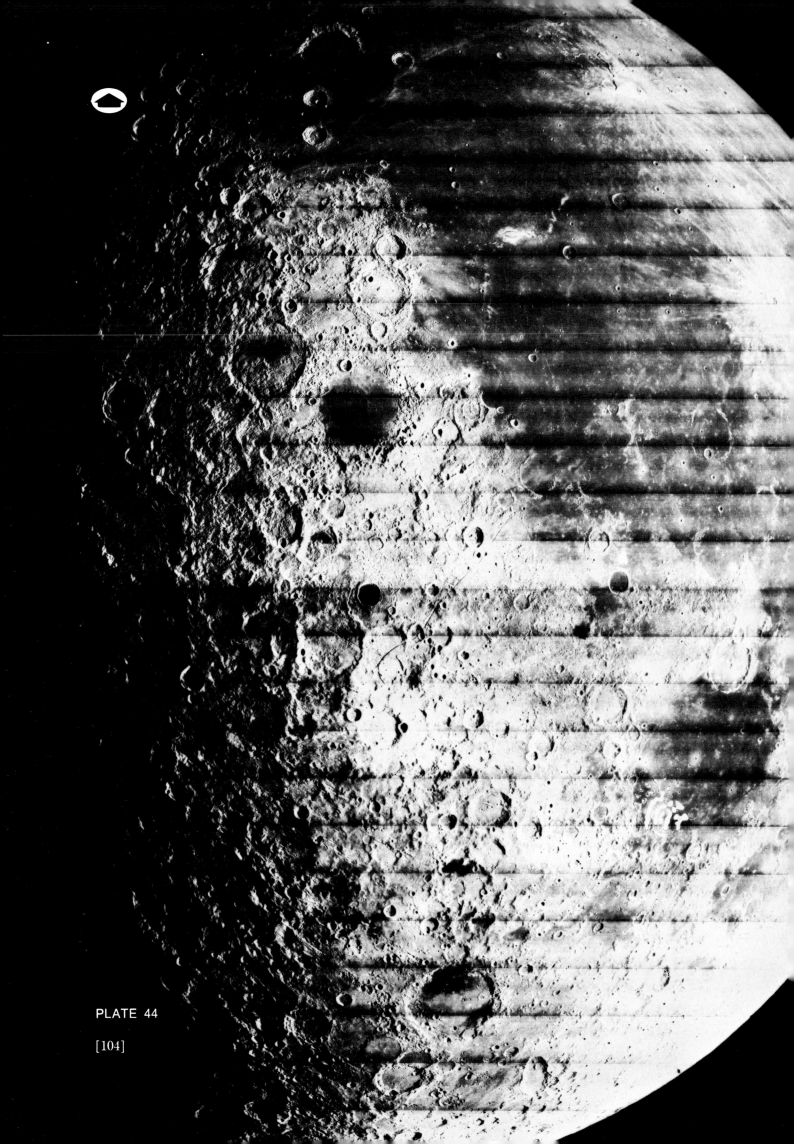

PLATE 44

[104]

45. Sunrise over the eastern ramparts of the lunar Mare Orientale: a huge triple-walled basin of marked circular symmetry, almost 900 km across, photographed by Orbiter 4 on May 24, 1967 from an altitude of 2,722 km. The trio of craters Eddington, Russell, and Struve is seen near the upper right of the frame; with Grimaldi and Riccioli below, and Schickard near the lower right corner. (Reproduced by courtesy of NASA-Langley and of the Boeing Company)

46. An oblique view of Mare Orientale photographed by Orbiter 4 on May 25, 1967 from an altitude of 2,673 km. The crater Einstein is at the center of the field, with Eddington, Russell, and Struve to the northeast and Roentgen above it. The dark-floor craters Grimaldi and Riccioli appear to the northeast of Mare Orientale; while the crater with a central mountain on the outer slopes of the Cordillera Mountains in Schlüter. (Reproduced by courtesy of NASA-Langley and of the Boeing Company)

47. "Bull's-eye" view of Mare Orientale, photographed by Orbiter 4 on May 25, 1967 from an altitude of 2,721 km. White rectangles delimit areas shown at higher resolution on Plates 48–50. (Reproduced by courtesy of NASA-Langley and of the Boeing Company)

48. High-resolution view of the northeast section of the "outer ramparts" of Mare Orientale, showing detailed structure of the ground; as photographed with the telephoto lens of Orbiter 4 on May 24, 1967 from an altitude of 2,722 km. (Reproduced by courtesy of NASA-Langley and of the Boeing Company)

49. A high-resolution view of a southeastern section of the outer ramparts of Mare Orientale, showing the structure of its outward-sloping walls. Photograph taken with the telephoto lens of Orbiter 4 on May 24, 1967 from an altitude of 3,007 km. (Reproduced by courtesy of NASA-Langley and of the Boeing Company)

50. A high-resolution view of a southern part of the inner floor of Mare Orientale and of its inner two rings, as photographed with the telephoto lens of Orbiter 4 on May 15, 1967 from an altitude of 2,719 km. (Reproduced by courtesy of NASA-Langley and of the Boeing Company)

51. Sunset over the central plains of Oceanus Procellarum between the craters Kepler (left) and Lansberg (at the sunset terminator), below the Carpathian Mountains. Photograph taken on September 20, 1965 with the 43-inch reflector at Observatoire du Pic-du-Midi. (Manchester Lunar Program)

52. Sunrise over the southwest plains of the Oceanus Procellarum. The craters at the extreme north (top) of the sunrise terminator are Cavalerius and Hevelius, already familiar from Plates 43 and 44. The two craters near the opposite corner of the plate are Billy and Hansteen. In the lower left-hand corner can be seen the well-known Sirsalis rille, shown in greater detail on Plate 181. Photograph taken on January 5, 1966 with the 43-inch reflector at Observatoire du Pic-du-Midi. (Manchester Lunar Program)

53. Sunrise over the crater Gassendi and Mare Humorum; to the east, Mare Nubium with the crater Bullialdus; to the south, Palus Epidemiarum. Photograph taken on July 8, 1965 with the 74-inch reflector of Helwan Observatory at Kottamia, Egypt. (Manchester Lunar Program)

PLATE 45

[106]

PLATE 46

PLATE 47

[108]

PLATE 8

[109]

PLATE 49

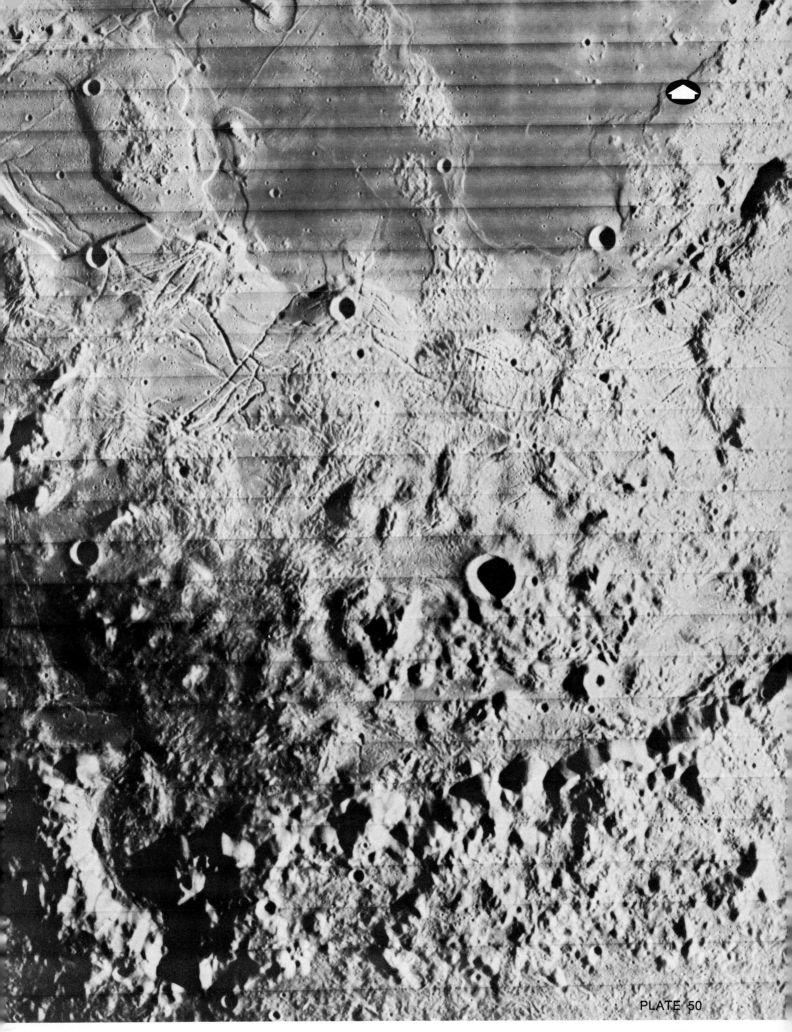

PLATE 50

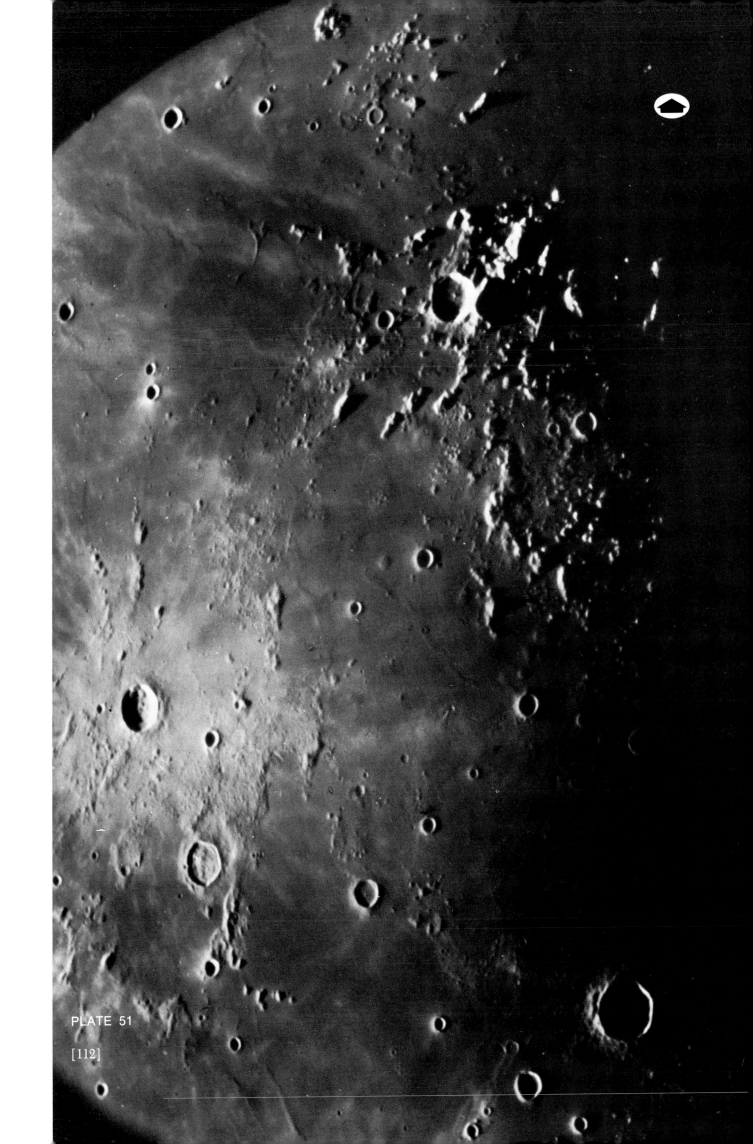

PLATE 51

[112]

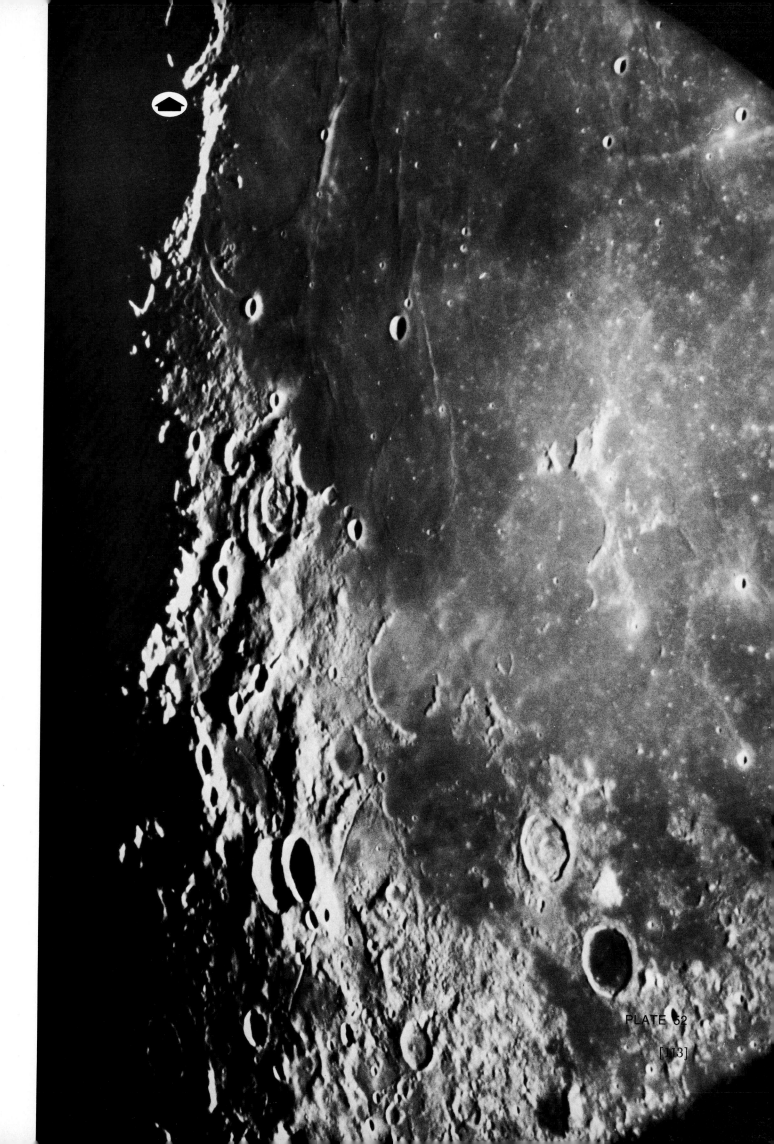

PLATE 52

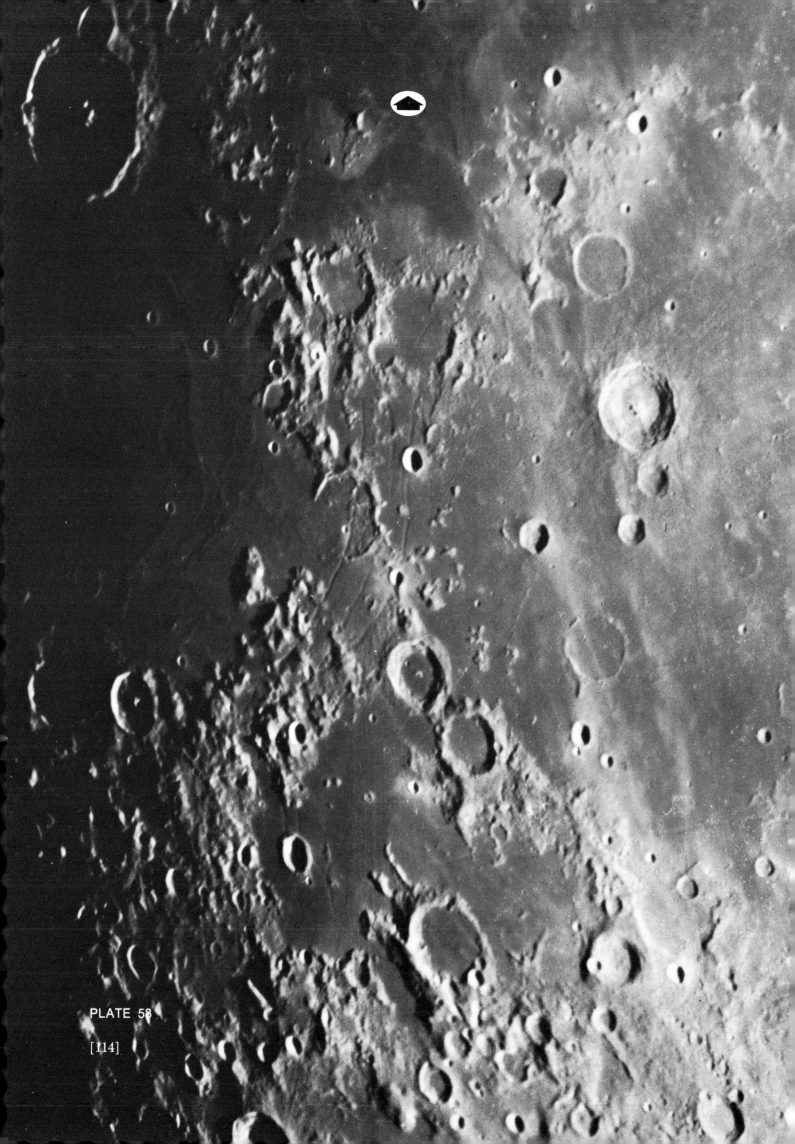

PLATE 58

[114]

54. Sunset over the lunar Mare Humorum (bordered by its characteristic system of rilles) and the crater Gassendi (extreme left). Photograph taken on September 20, 1965 with the 43-inch reflector at Observatoire du Pic-du-Midi. (Manchester Lunar Program)

55. Sunrise over Mare Humorum, with its system of rilles (right) and Cape Kelvin in the middle. Photograph taken on November 12, 1967 with the 43-inch reflector at Observatoire du Pic-du-Midi. (Manchester Lunar Program)

56. Mare Nubium, stretching between the Straight Wall to the south to the craters Lalande and Mösting to the north. Photograph taken on March 30, 1966 with the 43-inch reflector at Observatoire du Pic-du-Midi. (Manchester Lunar Program)

57. Sunrise over the southern part of Mare Nubium, between the craters Pitatus (bottom) and Guericke (top), with the Straight Wall in the middle. Photograph taken on March 30, 1966 with the 43-inch reflector at Observatoire du Pic-du-Midi. (Manchester Lunar Program)

58. Early sunrise over the Straight Wall in Mare Nubium, a fault approximately 90 km in length, and differing in height by less than 200 m (the actual slope of the "wall" nowhere exceeds 11°). The craters Alphonsus and Arzachel are visible near the top of the plate, and Deslandres is in the lower right corner. Photograph taken on August 24, 1966 with the 24-inch refractor at Observatoire du Pic-du-Midi. (Manchester Lunar Program)

59. Sunset over the Straight Wall in Mare Nubium (top) and the crater Deslandres (bottom). Photograph taken on August 9, 1966 with the 43-inch reflector at Observatoire du Pic-du-Midi. (Manchester Lunar Program)

60. Sunset over the lunar Mare Nectaris, with the crater Theophilus on its western shores, and Fracastorius to the south. Photograph taken on September 14, 1965 with the 43-inch telescope at Observatoire du Pic-du-Midi. (Manchester Lunar Program)

61. Late afternoon over the lunar Mare Tranquillitatis, between the craters Theophilus (bottom) and Janssen B (top). Photograph taken on March 30, 1966 with the 43-inch telescope at Observatoire du Pic-du-Midi. (Manchester Lunar Program)

62. An early sunset at the eastern limb of the lunar hemisphere visible from the Earth, featuring Mare Foecunditatis near the center of the field. A little to the northwest of its center we see the twin craters Messier with their characteristic twin bright streaks. The eastern component of this pair of craters can be seen at greater resolution on Plate 129. The large crater (with central peak) on the eastern shores of Mare Foecunditatis is Langrenus; and the one near its southern tip is Petavius; while Taruntius guards the straits between Mare Foecunditatis and Tranquillitatis (in the upper left corner). The southern part of Mare Crisium is visible on the top near the terminator; while Mare Nectaris is partly visible near the western (left) margin of the field. Photograph taken on November 21, 1964 with the 24-inch refractor of the Observatoire du Pic-du-Midi. (Manchester Lunar Program)

63. Sunset over the lunar Mare Crisium—a circular mare near the eastern limb of the lunar hemisphere visible from the Earth—photographed on January 7, 1966 with the 43-inch reflector at Observatoire du Pic-du-Midi. (Manchester Lunar Program)

PLATE 54

[116]

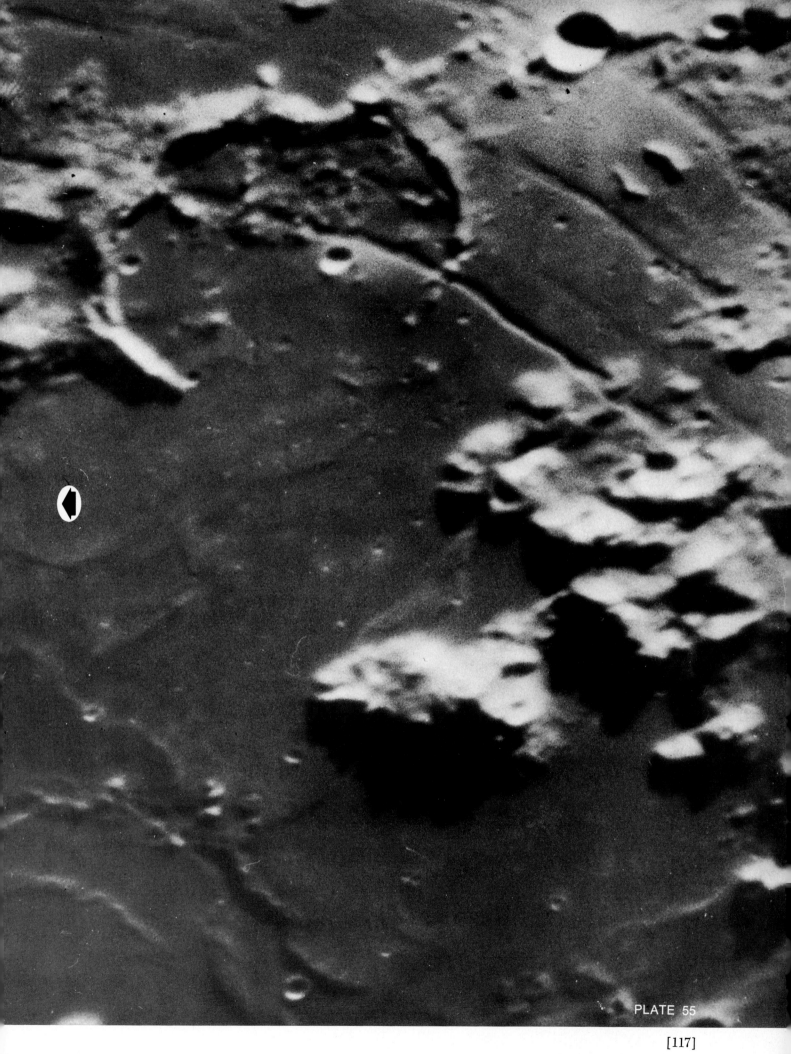

PLATE 55

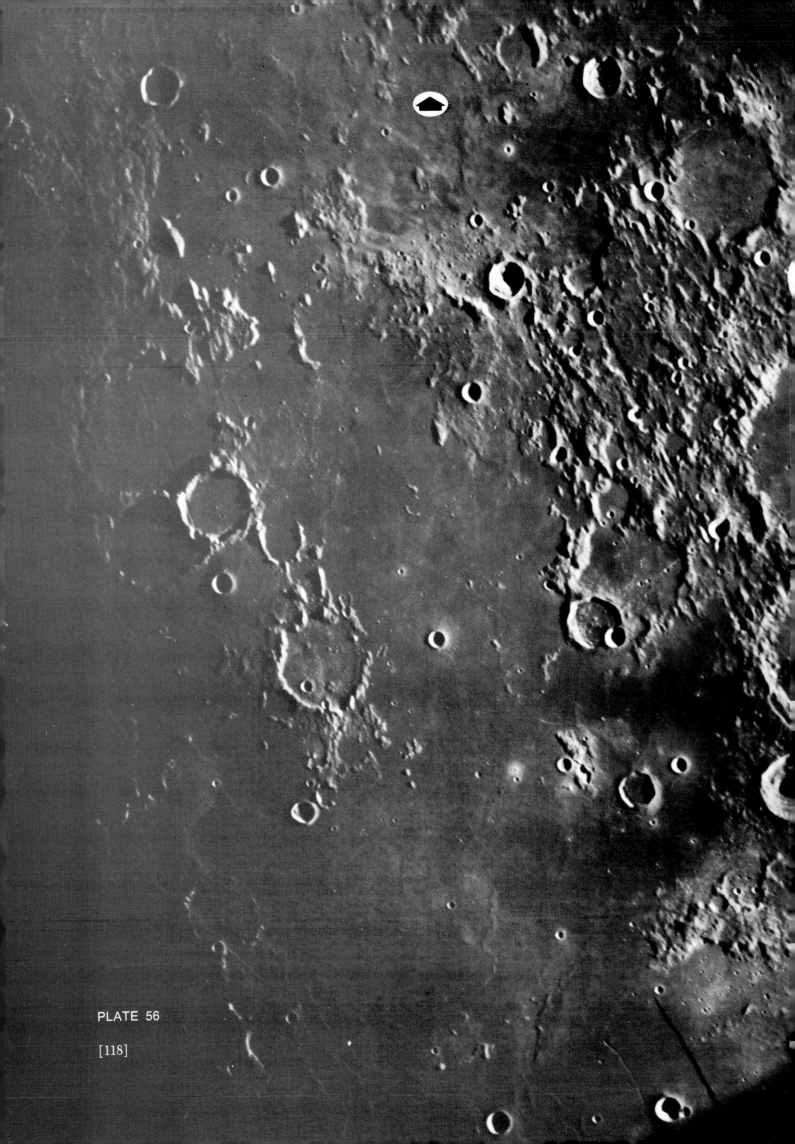

PLATE 56

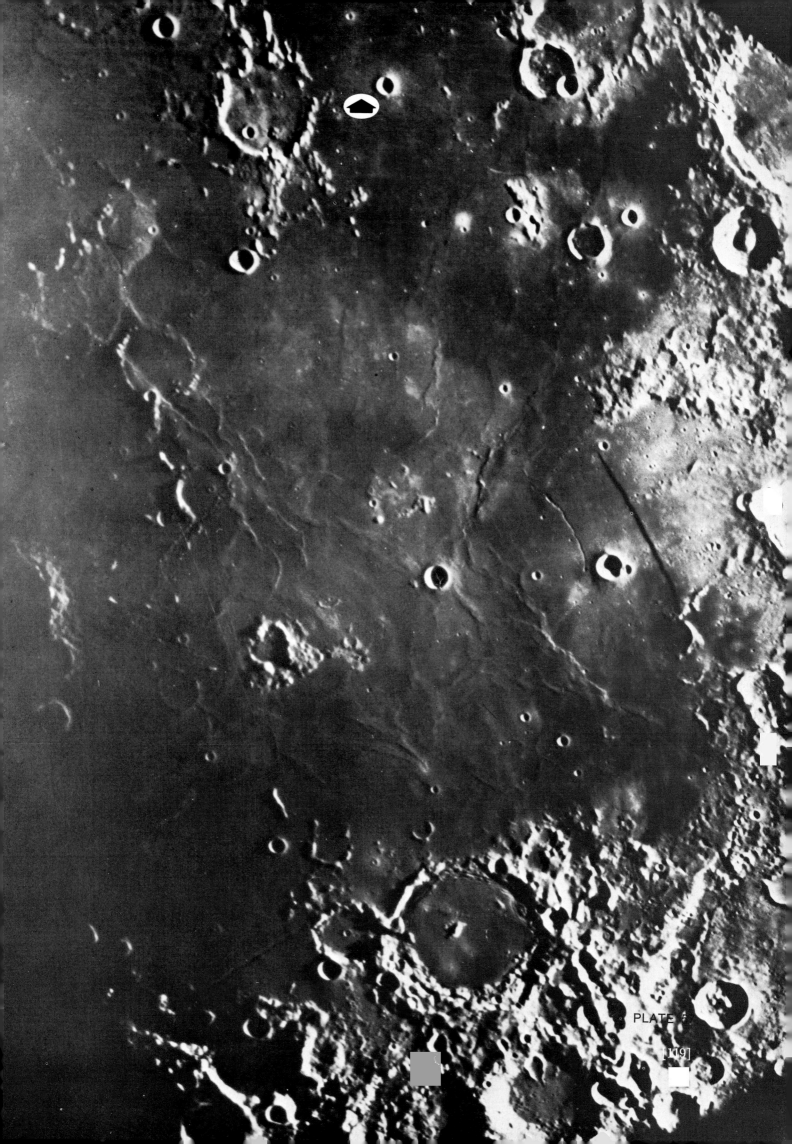

PLATE 6

[119]

PLATE 58

PLATE 59

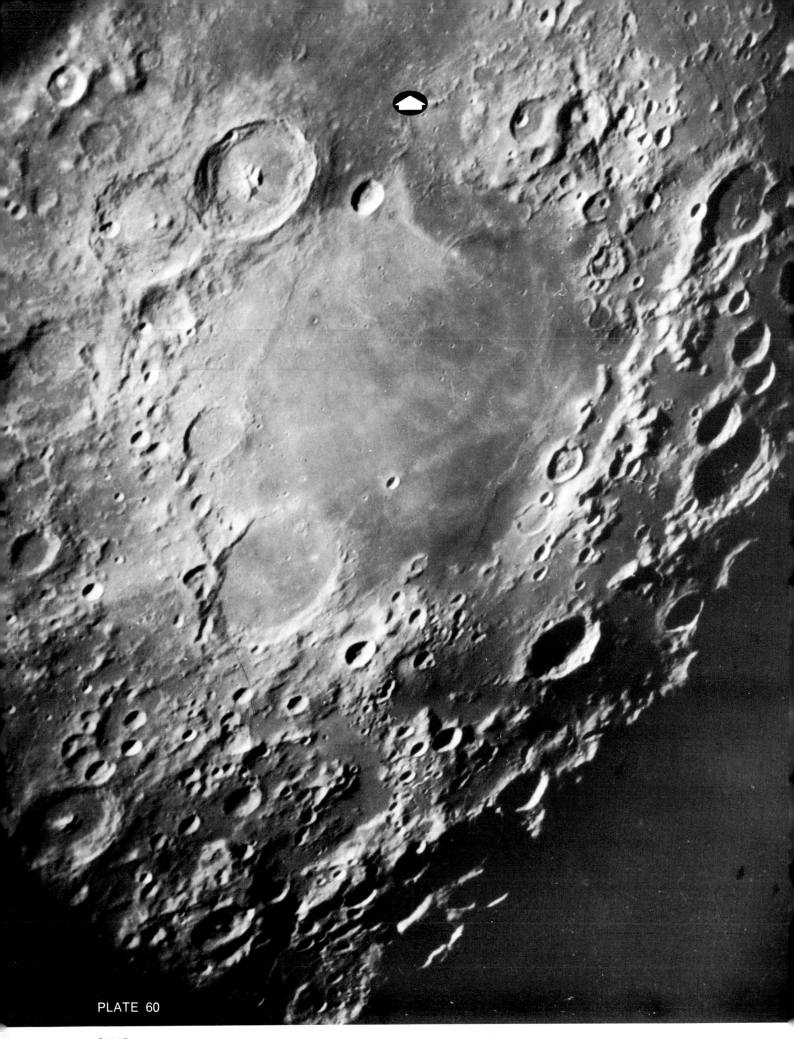

PLATE 60

PLATE 61

[123]

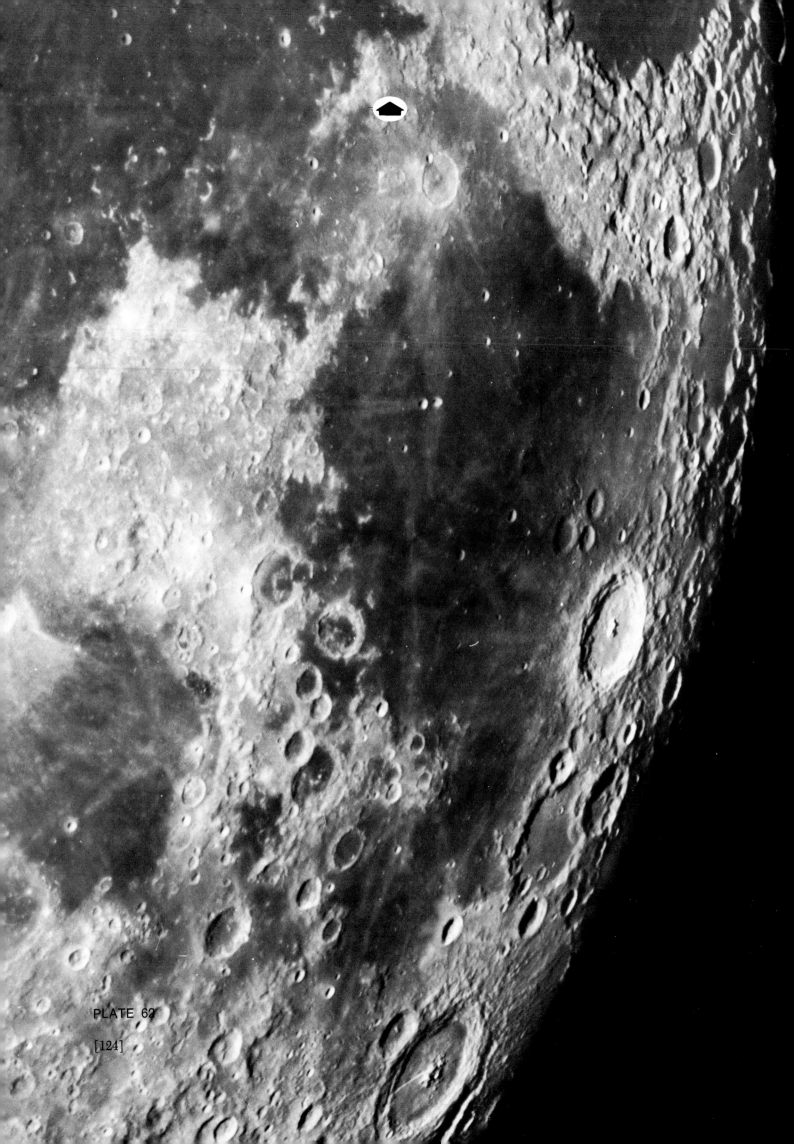

PLATE 62

[124]

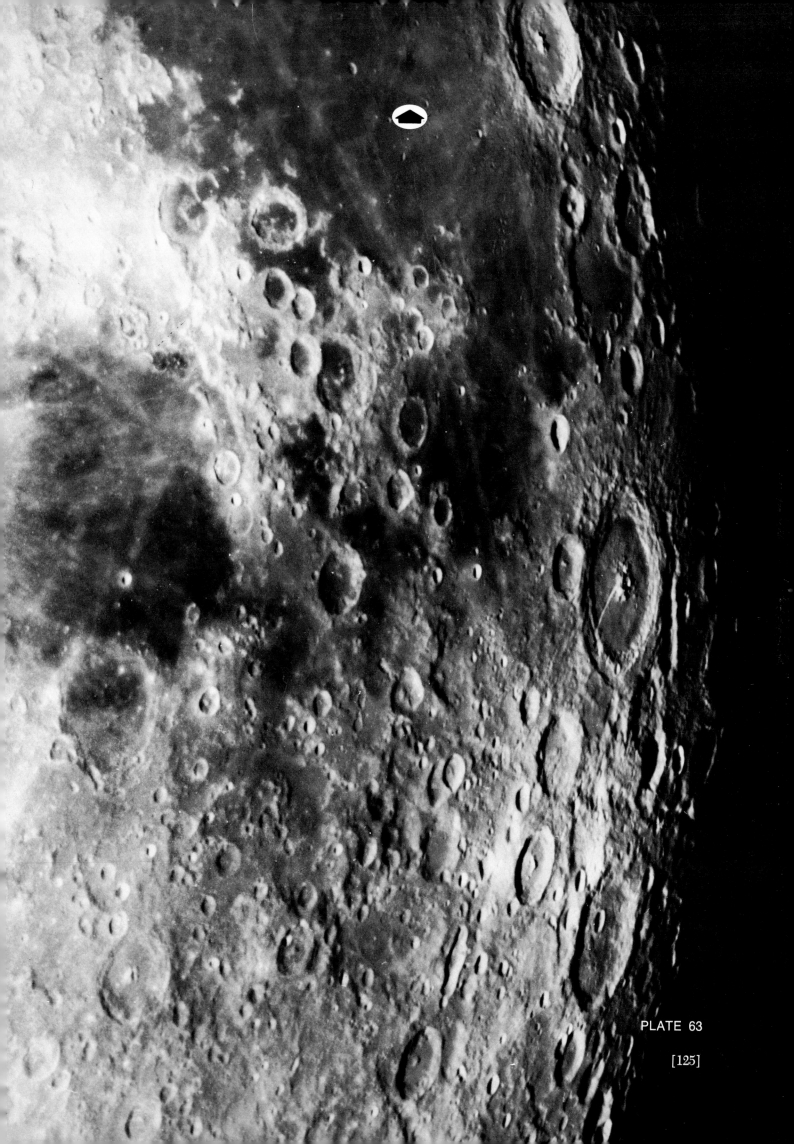

PLATE 63

[125]

64. Lunar Mare Tranquillitatis, from the crater Cauchy and its associated rille and cleft ("Cauchy's hyperbolae") on the upper left to Arago, Ross, and Plinius on the left. Photograph taken on November 23, 1964 with the 43-inch telescope at Observatoire du Pic-du-Midi. (Manchester Lunar Program)

65. Mare Vaporum and Sinus Medii near the center of the apparent disk of the Moon, photographed on September 16, 1965 with the 43-inch telescope at Observatoire du Pic-du-Midi. (Manchester Lunar Program)

66. Eastern shores of Mare Serenitatis dominated by the large crater Posidonius and the well-known Serpentine wrinkle ridge paralleling its eastern shores. Photograph taken on August 6, 1966 with the 43-inch reflector at Observatoire du Pic-du-Midi. (Manchester Lunar Program)

67. Lacus Somniorum between Posidonius and the twin craters Atlas and Hercules (upper right). Photographed on September 14, 1965 with the 43-inch reflector at Observatoire du Pic-du-Midi. (Manchester Lunar Program)

68. Sunset over the Lacus Somniorum and the twin craters Atlas and Hercules. Photograph taken on September 15, 1965 with the 43-inch reflector at Observatoire du Pic-du-Midi. (Manchester Lunar Program)

69. The crater Posidonius on the eastern shores of Mare Serenitatis, with the fork of the Serpentine wrinkle ridge to the west, photographed on November 24, 1964 with the 43-inch reflector at Observatoire du Pic-du-Midi. (Manchester Lunar Program)

70. Serpentine wrinkle ridge in Mare Serenitatis, photographed on November 24, 1964 with the 43-inch reflector at Observatoire du Pic-du-Midi. (Manchester Lunar Program)

71. Lacus Mortis and the eastern part of Mare Frigoris, between the craters Atlas and Hercules (right) and Aristoteles and Eudoxus (left). Photograph taken on November 23, 1964 with the 43-inch reflector at Observatoire du Pic-du-Midi. (Manchester Lunar Program)

72. Sunset over Lacus Mortis and the craters Aristoteles and Eudoxus south of Mare Frigoris, photographed on March 11, 1966 with the 43-inch reflector at Observatoire du Pic-du-Midi. (Manchester Lunar Program)

73. Mare Frigoris, north of the craters Aristoteles and Eudoxus, photographed on May 26, 1966 with the 43-inch reflector at Observatoire du Pic-du-Midi. (Manchester Lunar Program)

74. Western part of Mare Frigoris, north of Plato and west of the craters Archytas and Protagoras (see Plate 18) and Anaxagoras (with double central peak), photographed on July 9, 1966 with the 43-inch reflector at Observatoire du Pic-du-Midi. (Manchester Lunar Program)

75. Region near the north pole of the Moon, with the crater Pythagoras (one-half of which can be seen at the bottom), Carpenter (with double central peak), Anaximenes, Poncelet, and Pascal (upper left), as photographed on May 24, 1967 by the high-resolution lens of Lunar Orbiter 4 from an overhead vantage point at an altitude of 3,352 km above the lunar surface. Note the small crater Poncelet B, 16 km across (selenographic coordinates $\lambda = 53°2$ W, $\beta = 74°8$ N), marked with a white arrow near the top of the frame, the floor of which seems—like that of Wargentin—to be "filled to the brim." (Reproduced by courtesy of NASA-Langley and of the Boeing Company)

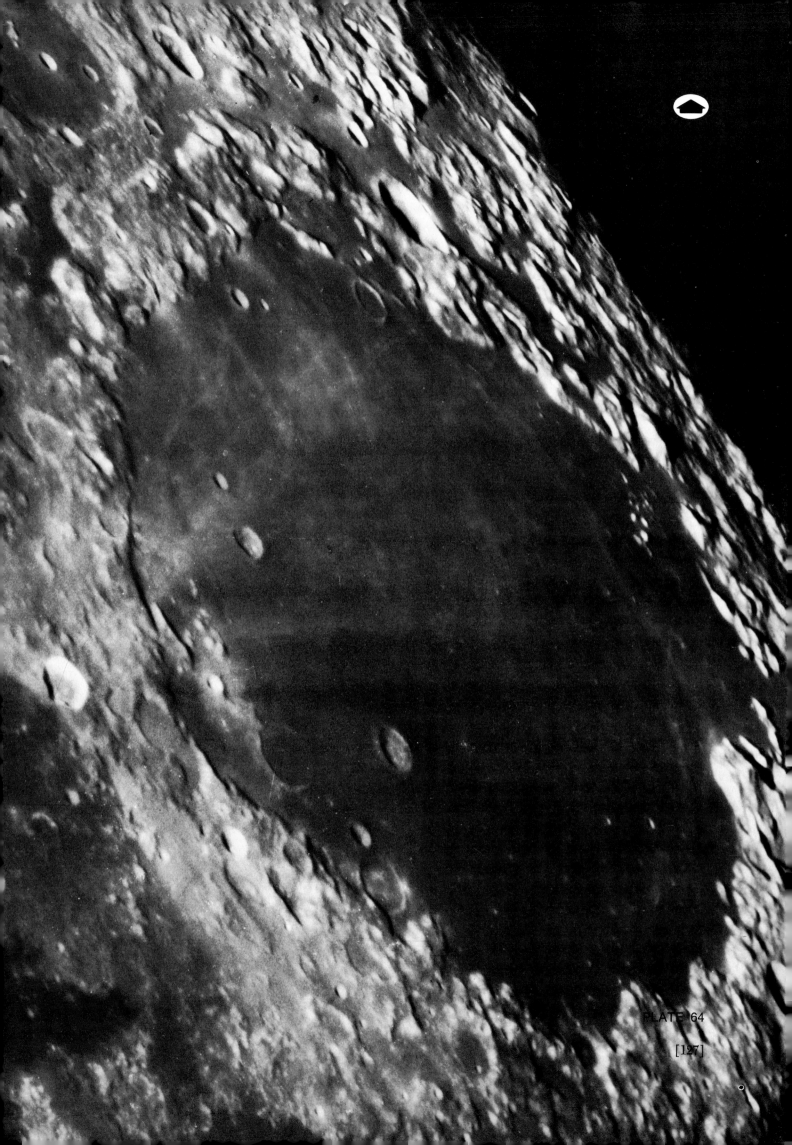

PLATE 64

[127]

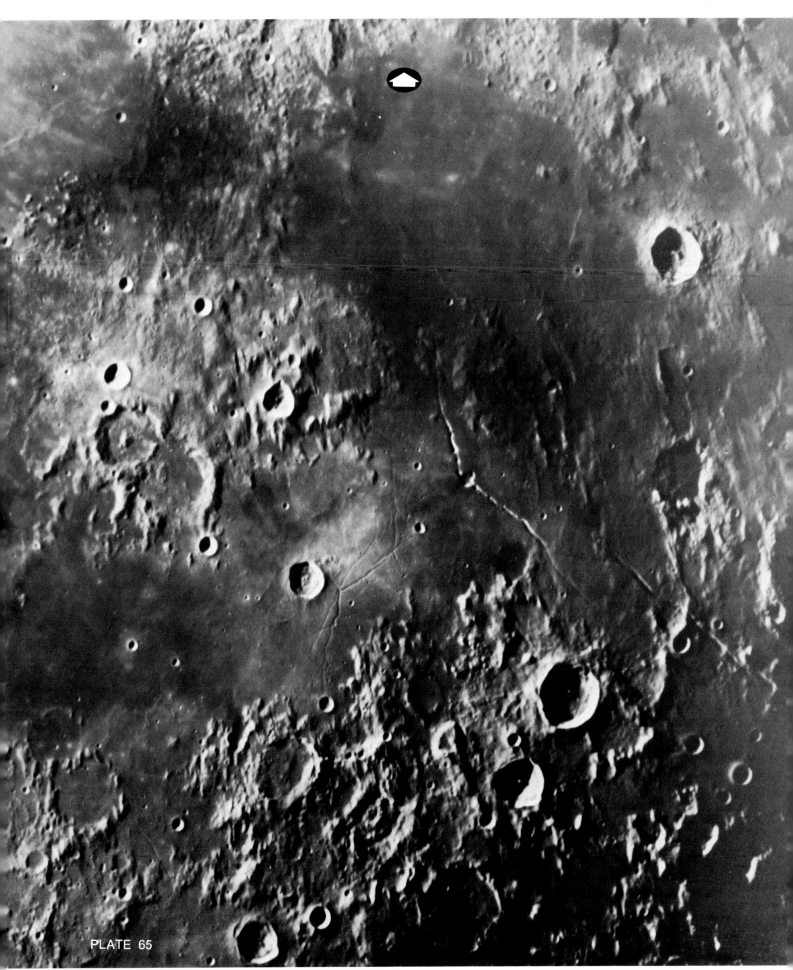

PLATE 65

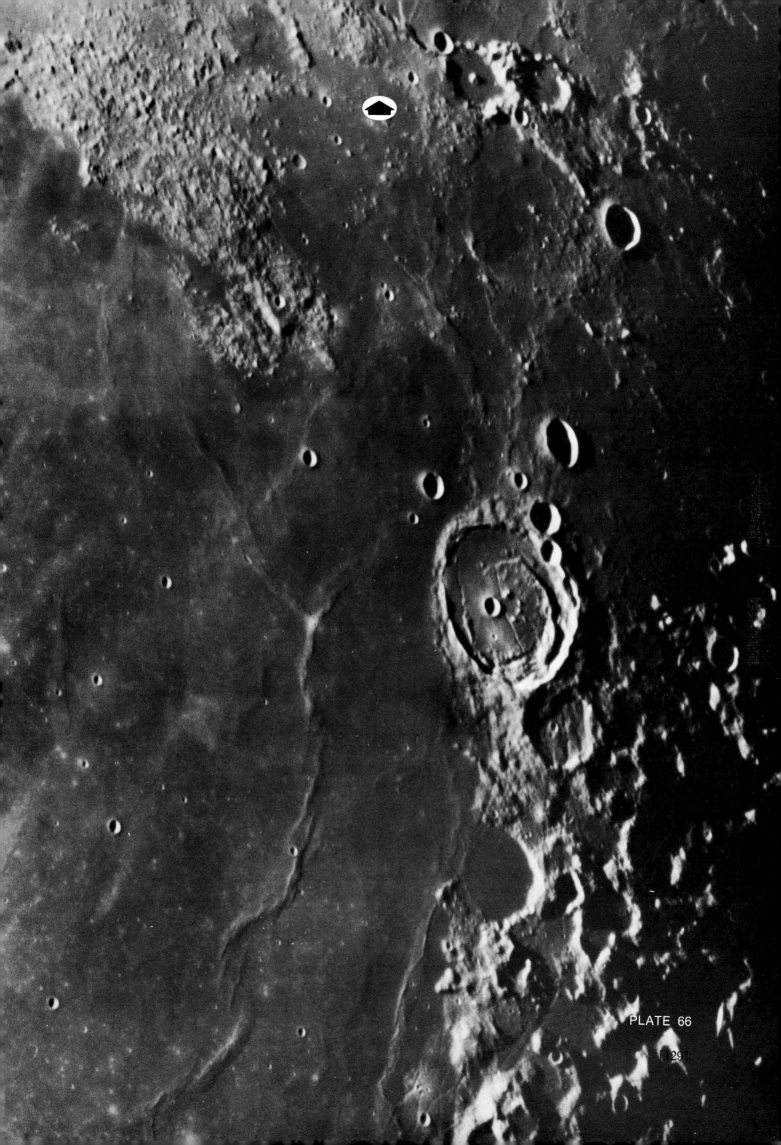

PLATE 66

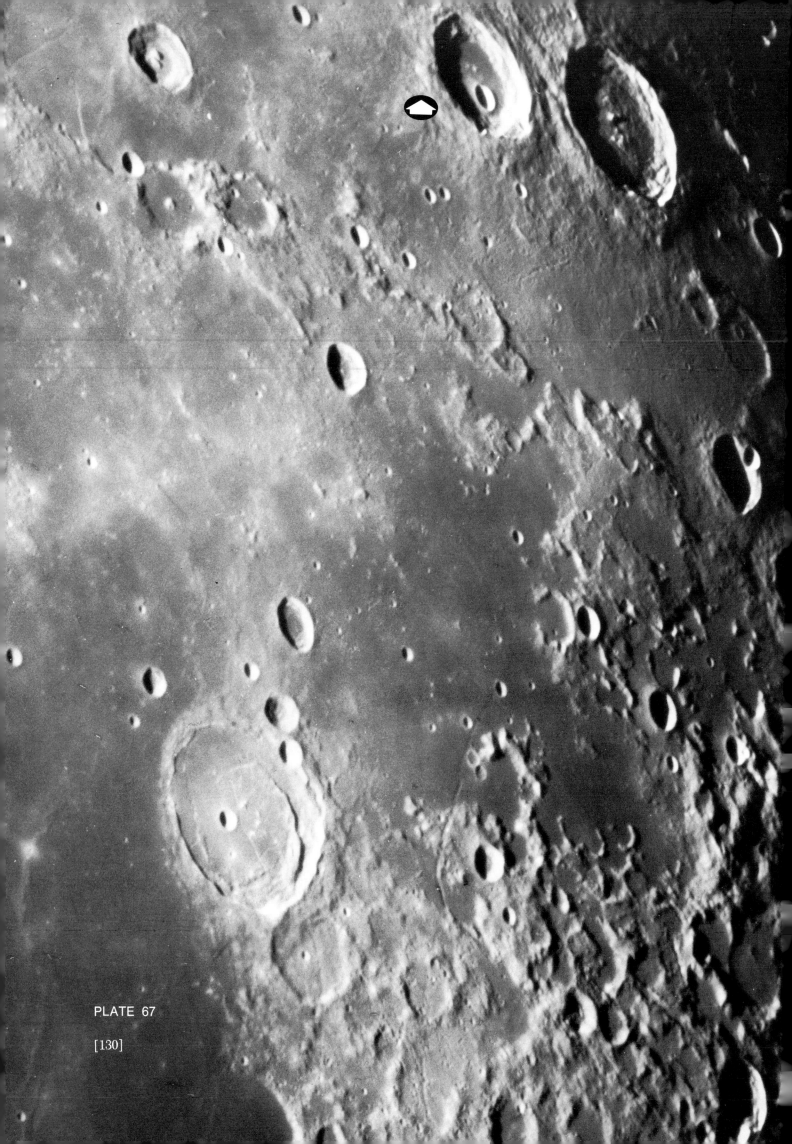

PLATE 67

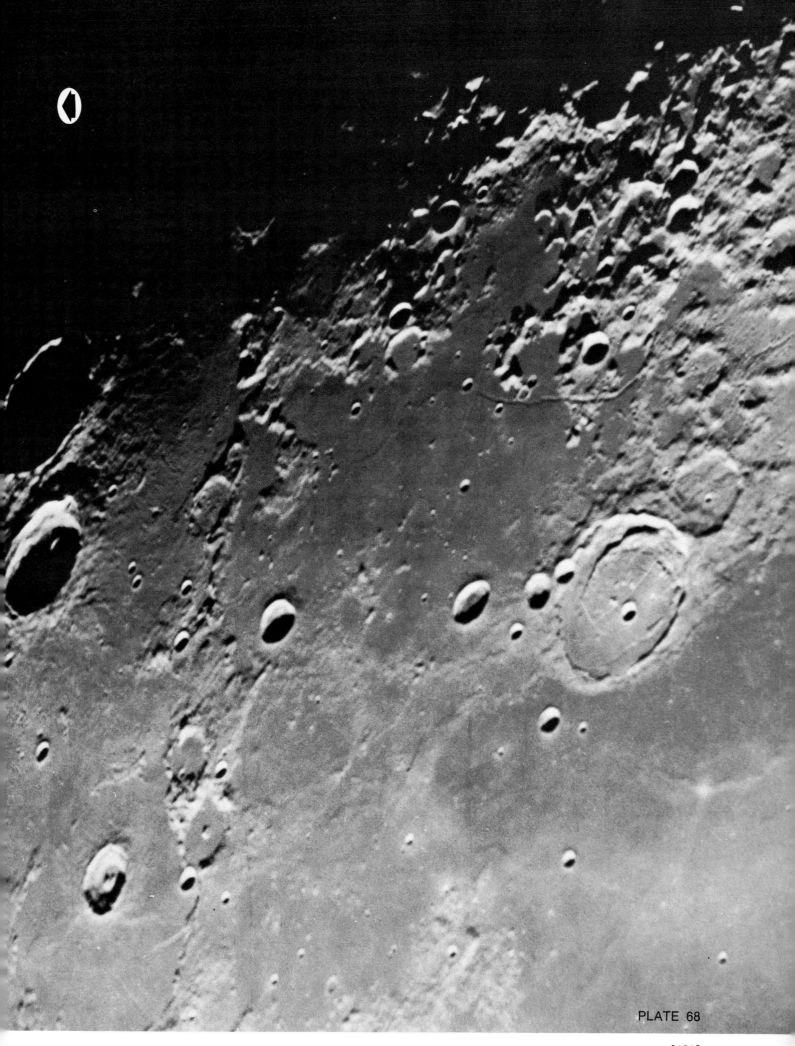

PLATE 68

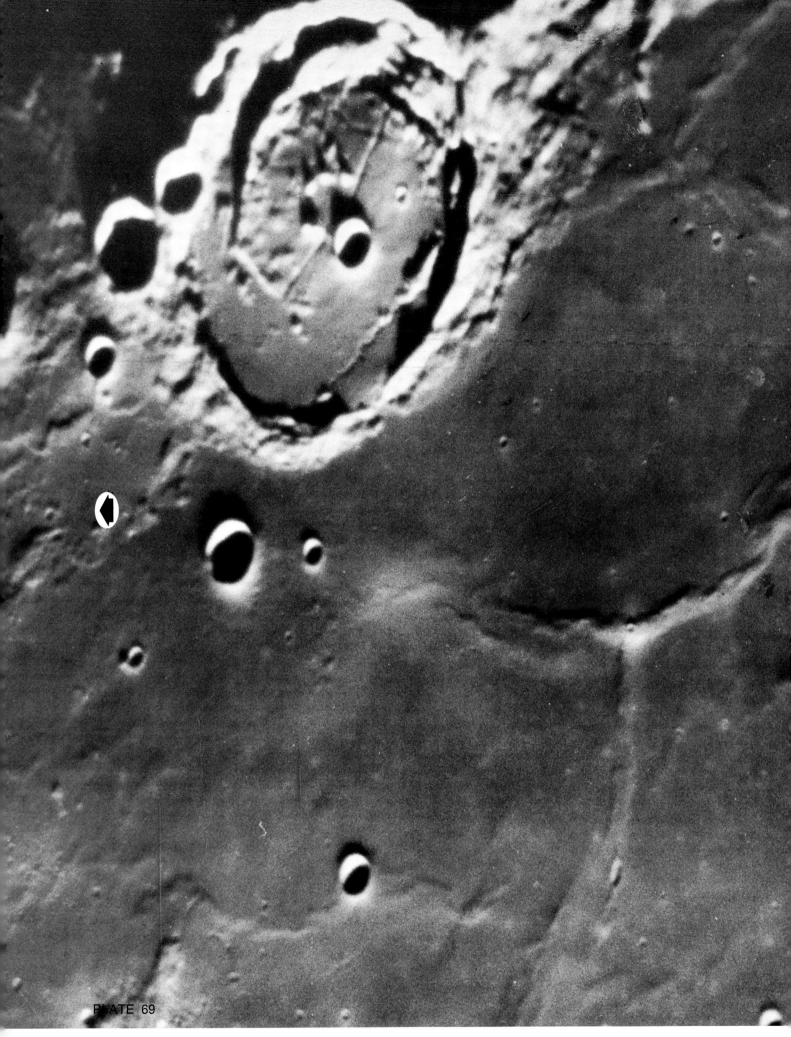

PLATE 69

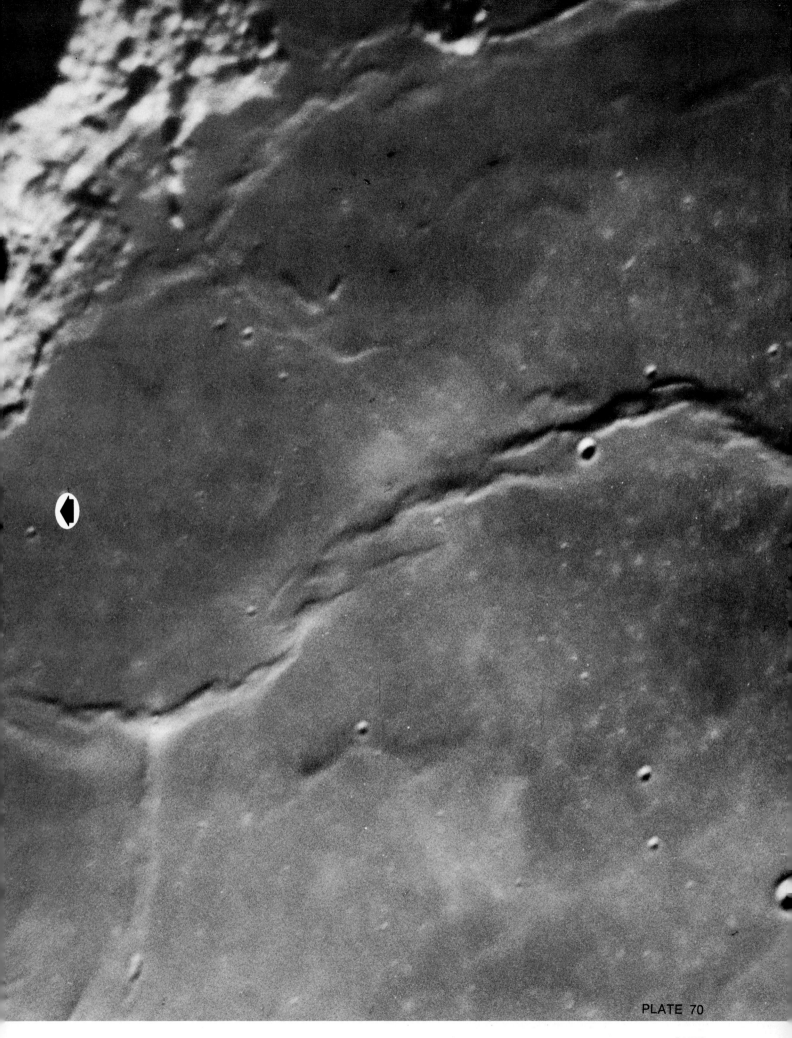

PLATE 70

[133]

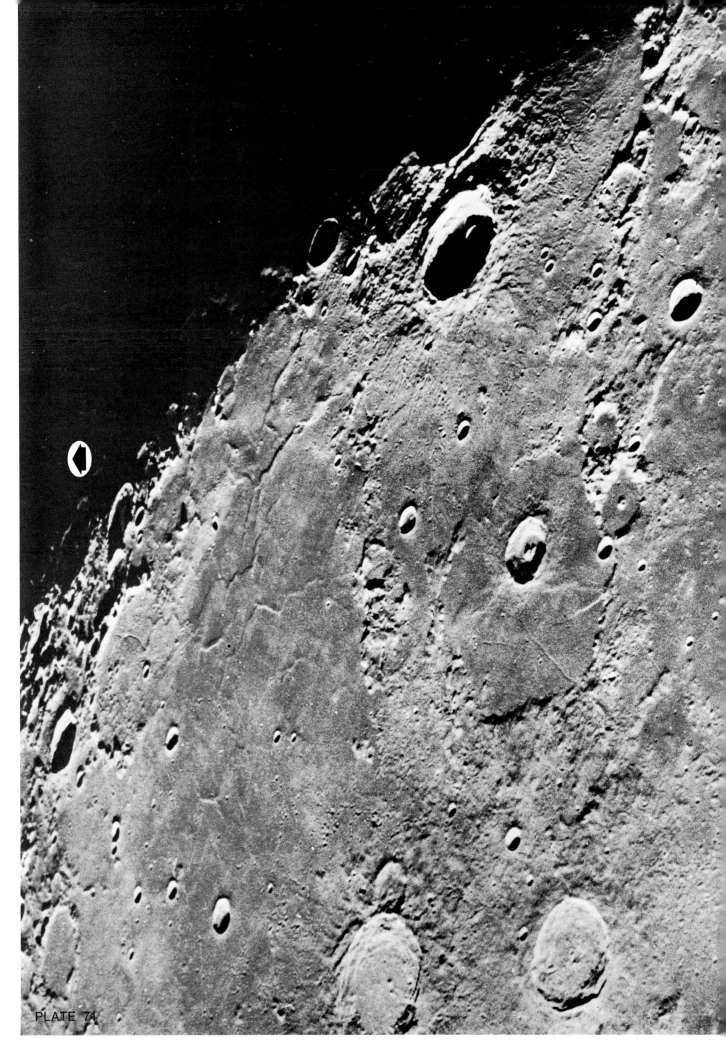

PLATE 71

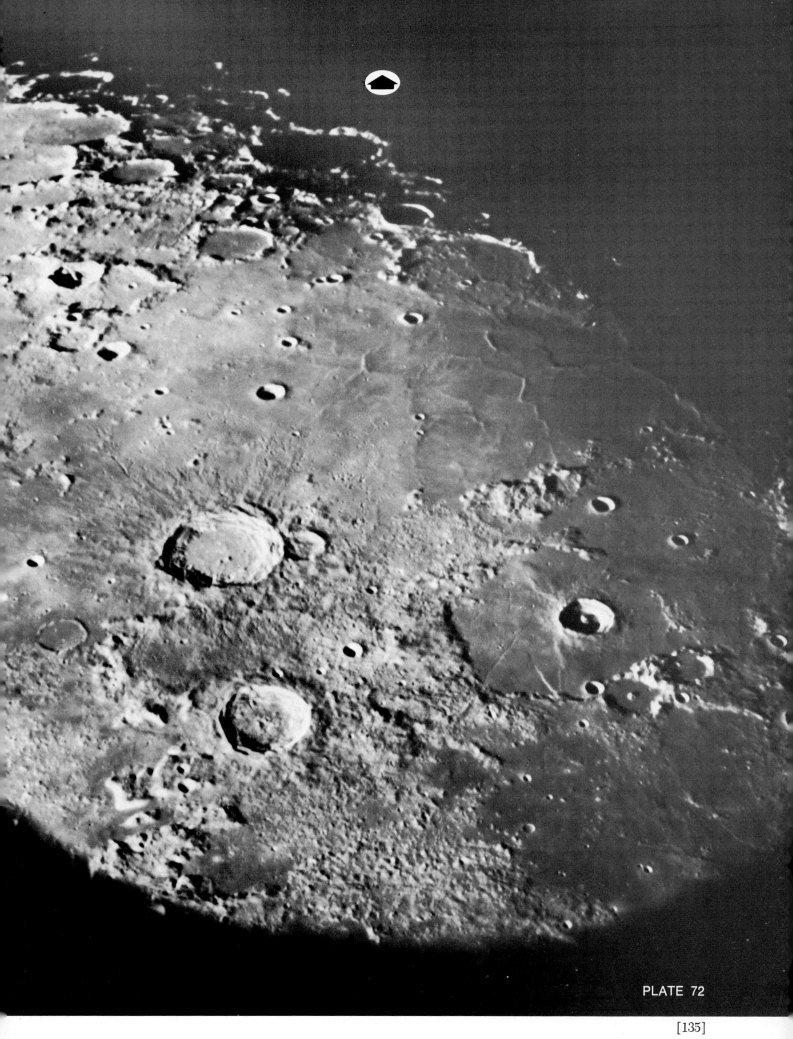

PLATE 72

PLATE 73

[136]

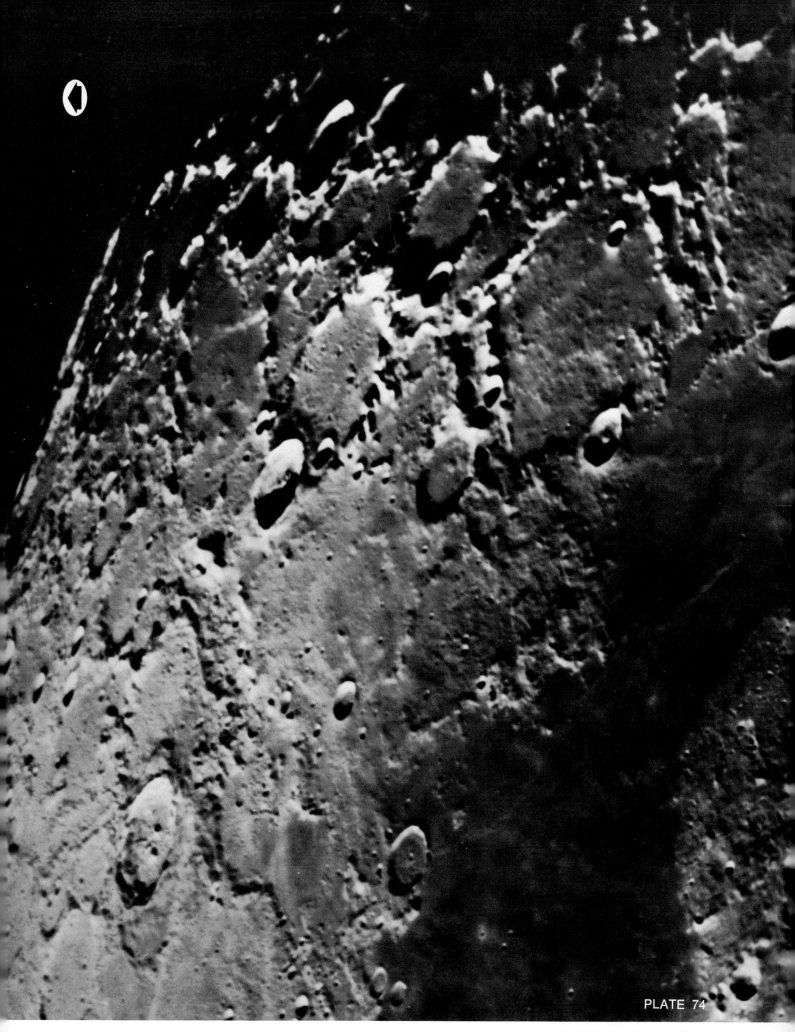

PLATE 74

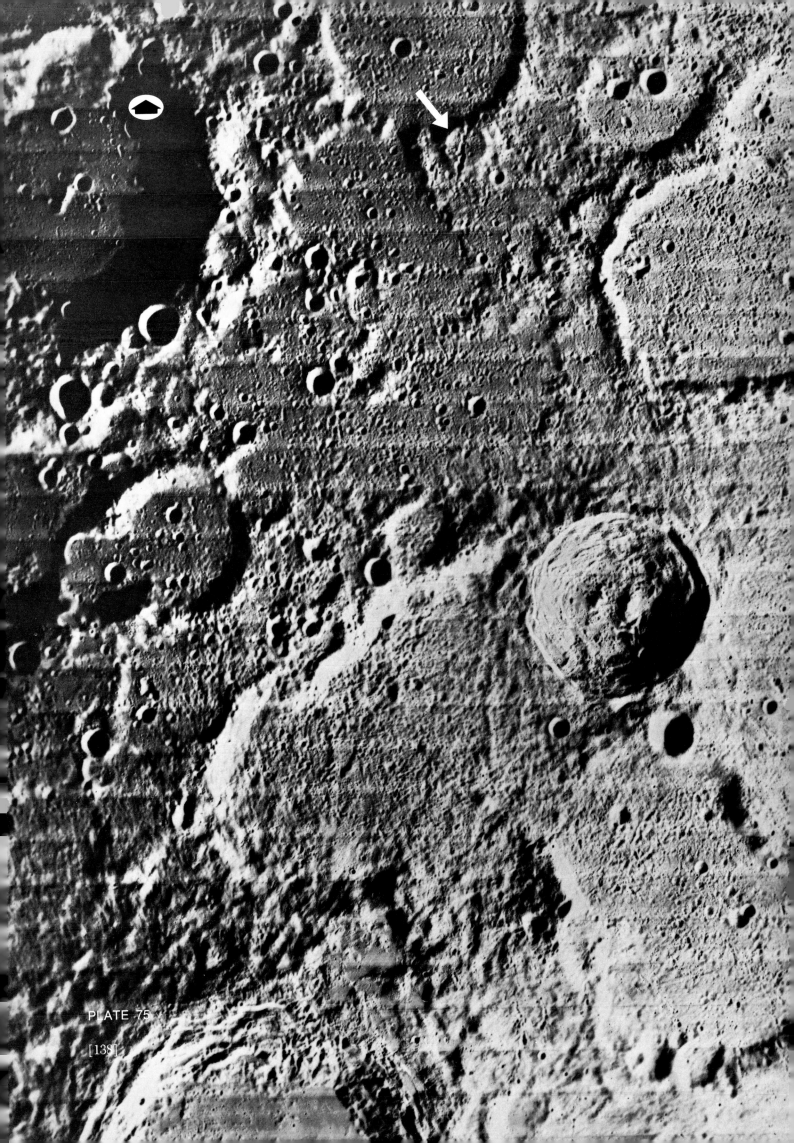

PLATE 75

[138]

76. Sunrise over a region adjacent to the lunar south pole, including the large craters Clavius and Maginus (to the left). Photograph taken on May 28, 1966 with the 43-inch reflector at Observatoire du Pic-du-Midi. (Manchester Lunar Program)

77. Sunrise over the craters Clavius (bottom), Maginus (center) and Tycho (upper left). Photograph taken on March 30, 1966 with the 43-inch reflector at Observatoire du Pic-du-Midi. (Manchester Lunar Program)

78. Sunset over the southern hemisphere of the Moon, between the crater Clavius (bottom) and the southern shores of Mare Imbrium with the crater Pitatus and the Hesiodus rille near the top. Only the eastern rim of Tycho (center) is still illuminated by the Sun; with Longomontanus to the left (west). Photograph taken on August 20, 1965 with the 43-inch reflector at Observatoire du Pic-du-Midi. (Manchester Lunar Program)

79. A photograph of the sunrise over the crater Tycho (87 km across) and its surroundings, taken on March 30, 1966 with the 43-inch reflector at Observatoire du Pic-du-Midi. (Manchester Lunar Program)

80. A photograph of the crater Tycho, taken by Lunar Orbiter 5 on August 15, 1967 from an altitude of 215 km above the lunar surface. White rectangle marks the floor area shown at higher resolution on Plate 81. (Reproduced by courtesy of NASA-Langley and of the Boeing Company)

81. A high-resolution photograph of a section (approximately 14 × 14 km in size) of the floor of the crater Tycho, taken by Lunar Orbiter 5 on August 15, 1967 from an altitude of 217 km above the lunar surface. (Reproduced by courtesy of NASA-Langley and of the Boeing Company)

82. Sunset over the crater Schiller (elongated, below) and its surroundings on the southwest limb of the Moon. Photograph taken on September 20, 1965 with the 43-inch reflector at Observatoire du Pic-du-Midi. (Manchester Lunar Program)

83. Southwestern limb of the visible lunar hemisphere, dominated by the large craters Schickard (center) and Schiller (elongated formation on the right). The outlines at sunrise of the peculiar crater Wargentin can be seen immediately below Schickard. Photograph taken on January 5, 1966 with the 43-inch reflector at Observatoire du Pic-du-Midi. (Manchester Lunar Program)

84. Sunrise over the large crater Schickard near the southwest limb of the visible lunar hemisphere. The enigmatic crater Wargentin is seen immediately to the west of Schickard, and above Phocylides; while the crater Schiller (see Plate 82) is half seen near the lower right corner. Photograph taken with the 120-inch reflector of the Lick Observatory in 1963.

85. Sunset over the crater Maurolycus (bottom); Stöfler, Walter, Regiomontanus, Purbach, and Arzachel to the left; Mare Nubium in the upper right corner. Photograph taken on August 18, 1965 with the 43-inch reflector at Observatoire du Pic-du-Midi. (Manchester Lunar Program)

86. A photograph of the craters Walter and Regiomontanus (Deslandres to the left, see also Plate 59; Werner and Aliacensis to the right) taken on May 18, 1967 by Lunar Orbiter 4 from an altitude of 2,980 km. Note the small formation Regiomontanus A (marked with an arrow) which may be of volcanic origin. (Reproduced by courtesy of NASA-Langley and of the Boeing Company)

PLATE 76

[140]

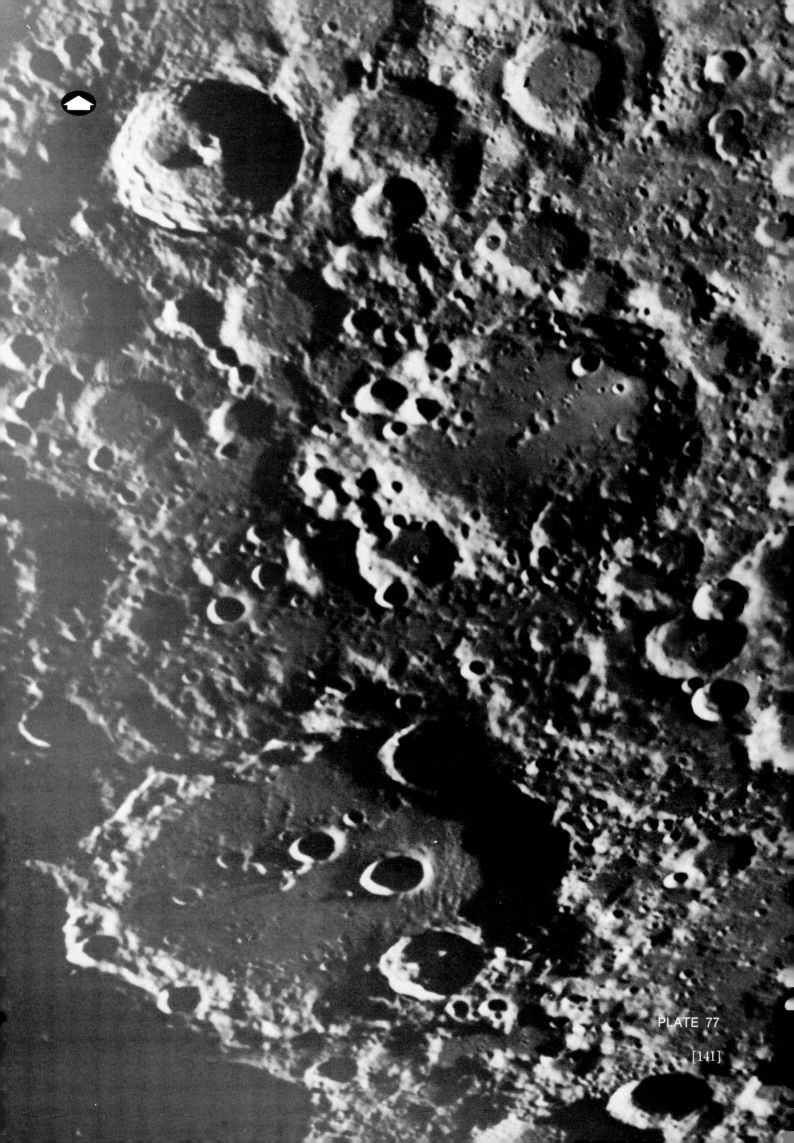

PLATE 77

PLATE 78

[142]

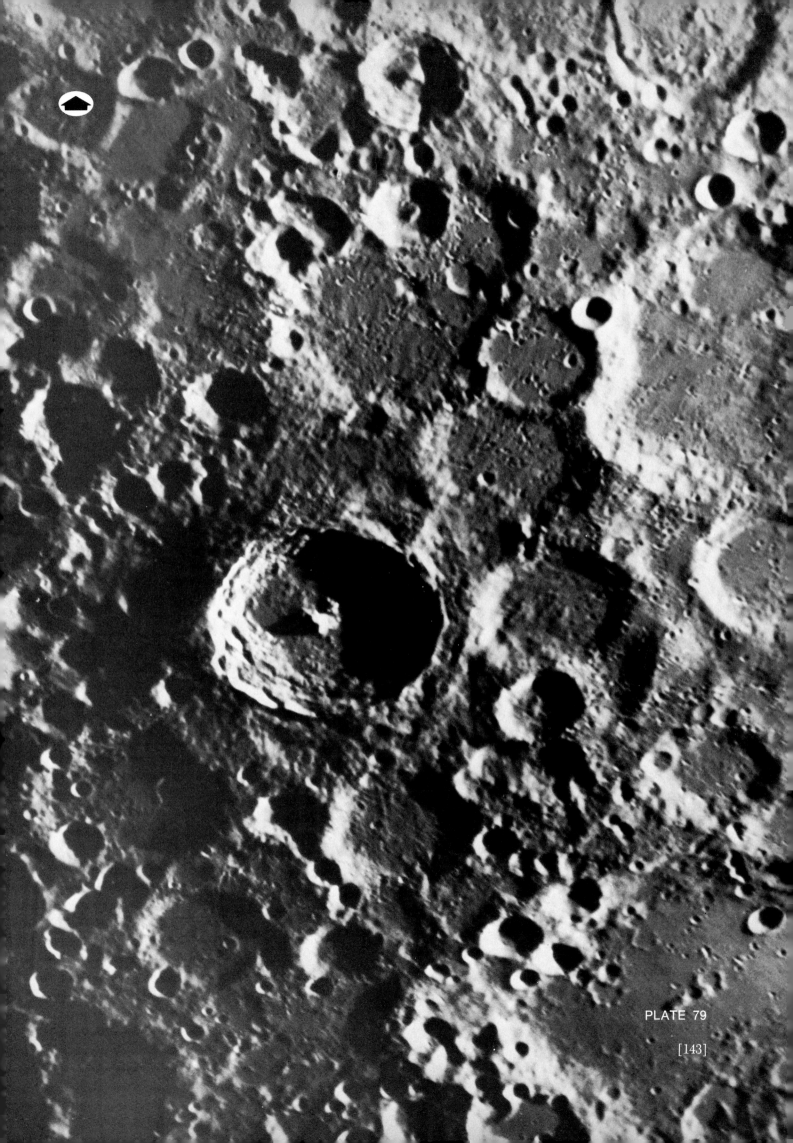

PLATE 79

PLATE 81

PLATE 82

[146]

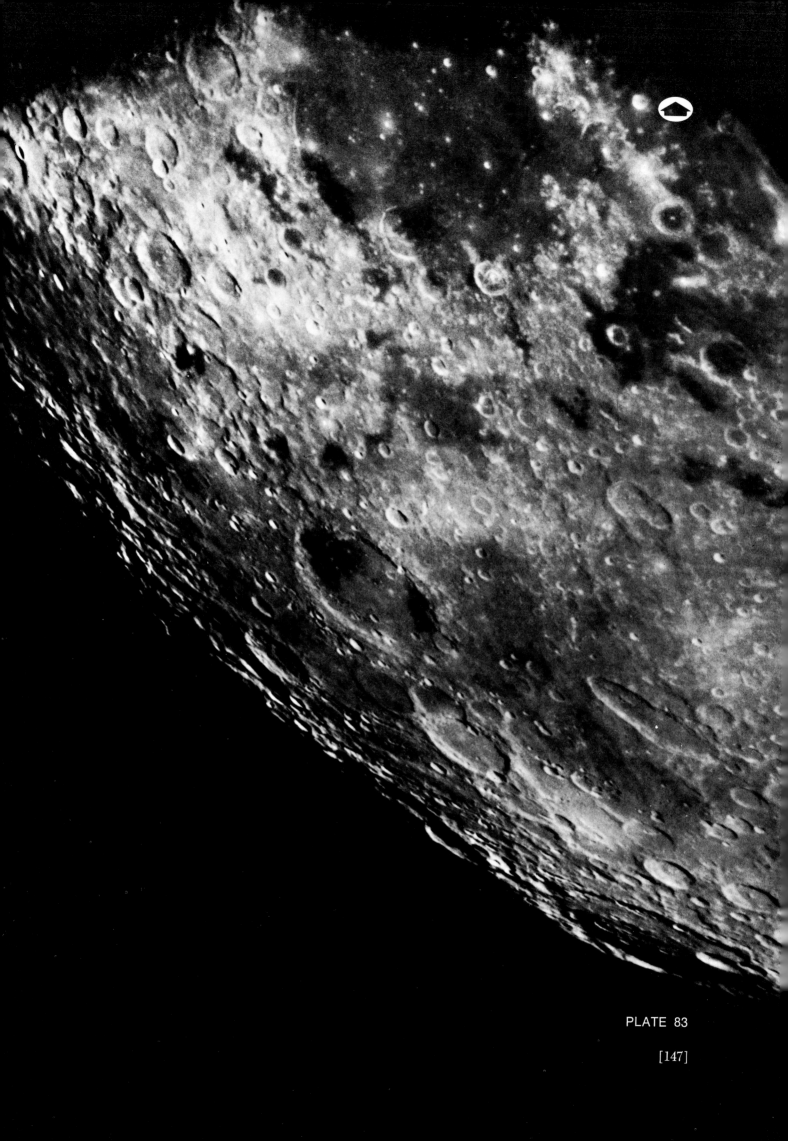

PLATE 83

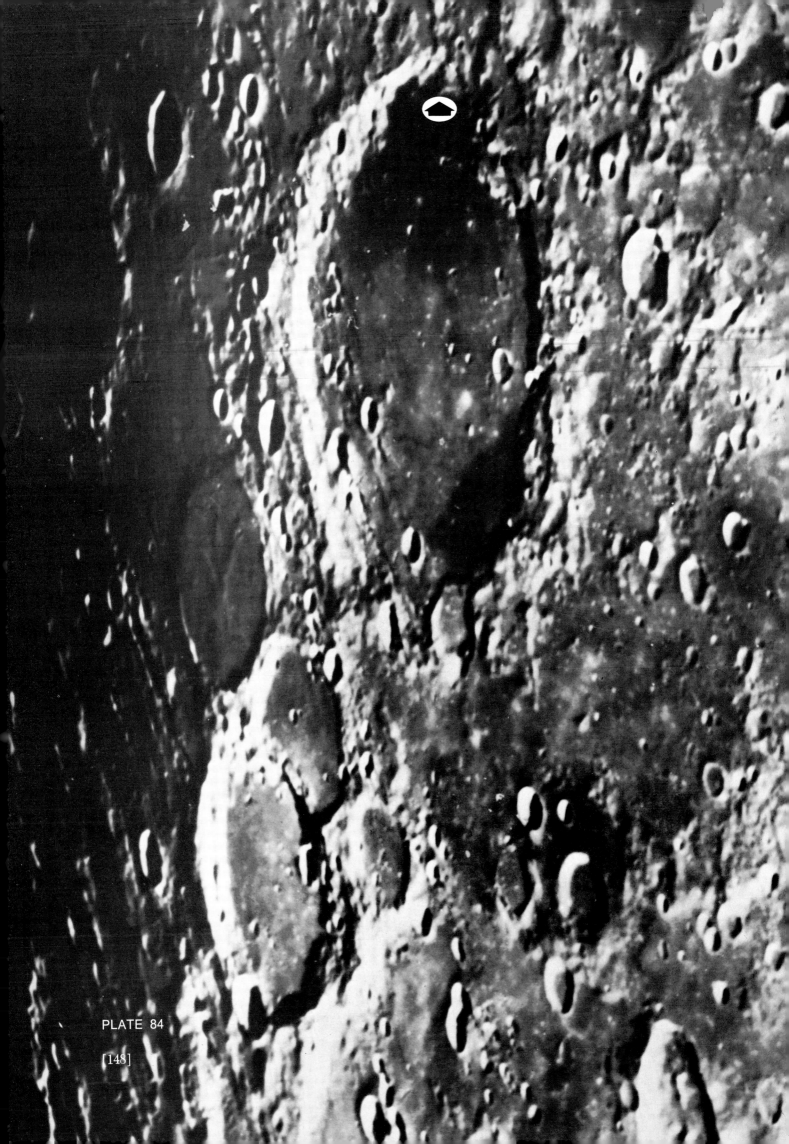

PLATE 84

[148]

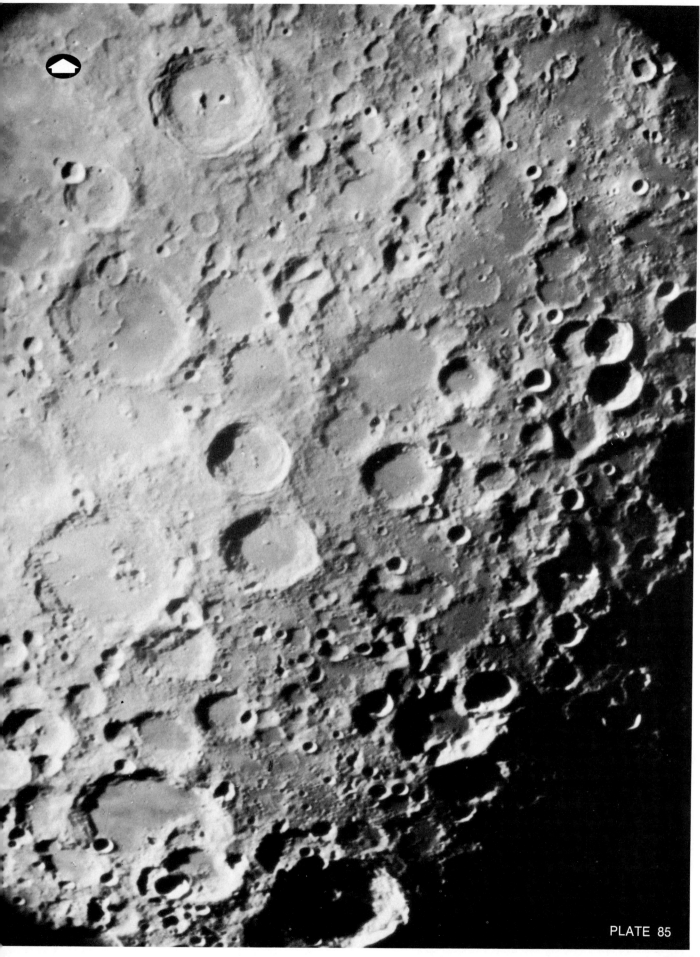

PLATE 85

PLATE 86

87.   The craters Ptolemaeus, Alphonsus, and Arzachel on the eastern shores of Mare Nubium; with Albategnius near the sunset terminator. Several grooves diverging radially from Mare Imbrium (to the south) can be clearly seen on this picture. Photograph taken on September 17, 1965 with the 43-inch reflector at Observatoire du Pic-du-Midi. (Manchester Lunar Program)

88.   Sunrise over the craters Ptolemaeus, Alphonsus, and Arzachel (with Albategnius to the left), photographed on May 27, 1966 with the 43-inch reflector at Observatoire du Pic-du-Midi. (Manchester Lunar Program)

89.   Sunset over the craters Ptolemaeus, Alphonsus, and Arzachel, photographed on February 2, 1967 with the 24-inch refractor at Observatoire du Pic-du-Midi. (Manchester Lunar Program)

90.   An oblique view of the craters Ptolemaeus (above), Alphonsus (to the left), and Albategnius (to the right), televised by Ranger 9's B-camera on March 24, 1965 from an altitude of 428 km above the lunar surface. (Reproduced by courtesy of NASA and of JPL)

91.   A view of the craters Alphonsus and Alpetragius, televised by Ranger 9's A-camera on March 24, 1965 from an altitude of 413 km above the lunar surface—2 min 50 sec before impact—recording a square field of 180 km in size. The plains of Mare Nubium on the left side of the frame are at a higher level, and more sparingly checkered by craters than the floor of the crater Alphonsus which obviously represents a very old type of lunar surface. Note also the rilles paralleling roughly the hexagonal outlines of the crater walls. (Reproduced by courtesy of NASA and of JPL)

92.   Mare-like formations on the floor of the crater Alphonsus, as televised by Ranger 9's B-camera on March 24, 1965 from an altitude of 184 km above the lunar surface. Note the greatly diminished reflectivity of the ground in the immediate surroundings of the mare formations, probably due to deposit of dark material which has come out of them—a view supported by the fact that the rille on which the mare lies is partly filled by ejecta on both sides of the crater. A part of the ramparts of Alphonsus can be seen to the right of the rilles. Note the enclosure within the walls (marked by a white arrow). Its ground bears striking resemblance to the floor of the crater, with which it may be generically related. (Reproduced by courtesy of NASA and of JPL)

93.   A view of the central part of the crater Alphonsus dominated by its "central mountain" and a shallow ridge running through it in the direction of the meridian. The surface of this ridge is much less disfigured by small craters —a fact suggesting later origin. Image televised by the A-camera of Ranger 9 on March 24, 1965 from an altitude of 166 km above the lunar surface, 1 min 9.5 sec before impact. (Reproduced by courtesy of NASA and of JPL)

94.   Sunrise over the craters Hipparchus and Albategnius (to the east of Ptolemaeus), photographed on August 4, 1965 with the 43-inch reflector at Observatoire du Pic-du-Midi. (Manchester Lunar Program)

95.   Sunrise over the craters Maurolycus and Barocius (east of Stöfler); with the Sun just rising over Cuvier and Licetus (left), photographed on May 26, 1966 with the 43-inch reflector at Observatoire du Pic-du-Midi. (Manchester Lunar Program)

96.   Sunrise over the crater Albategnius (left); a section of Hipparchus in the upper left corner; photographed on August 4, 1965 with the 43-inch reflector at Observatoire du Pic-du-Midi. (Manchester Lunar Program)

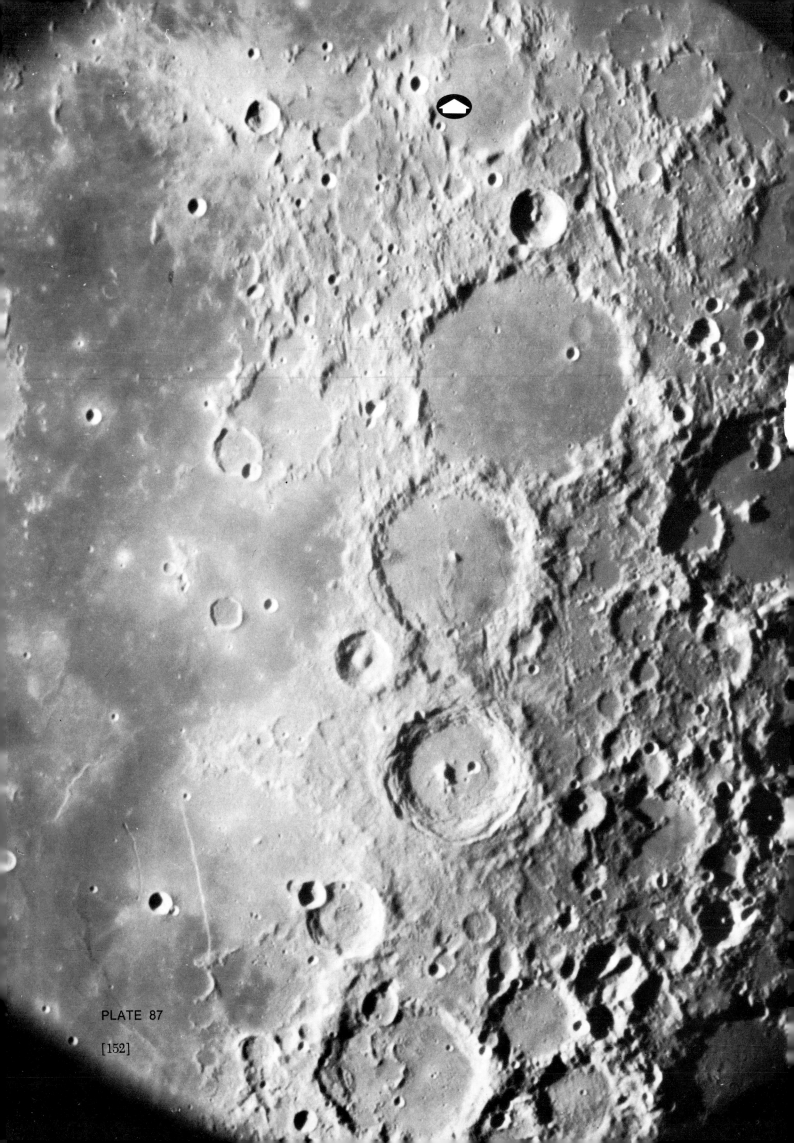

PLATE 87

[152]

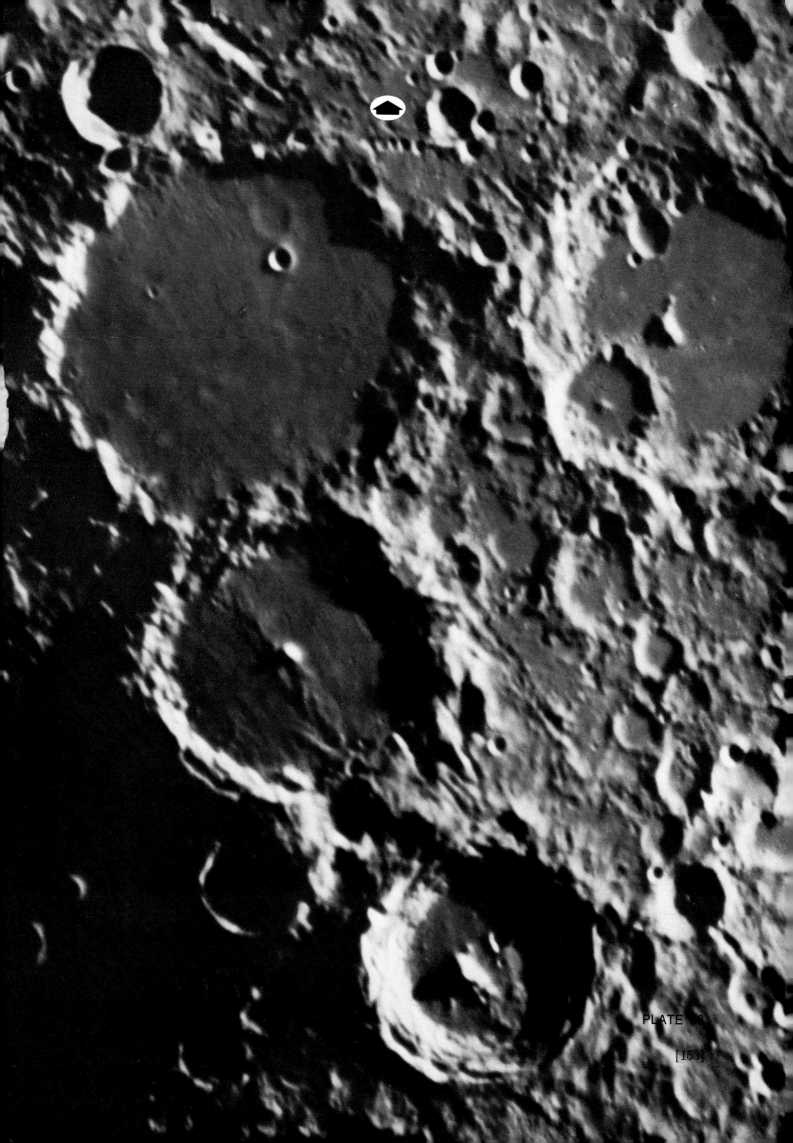

PLATE

[153]

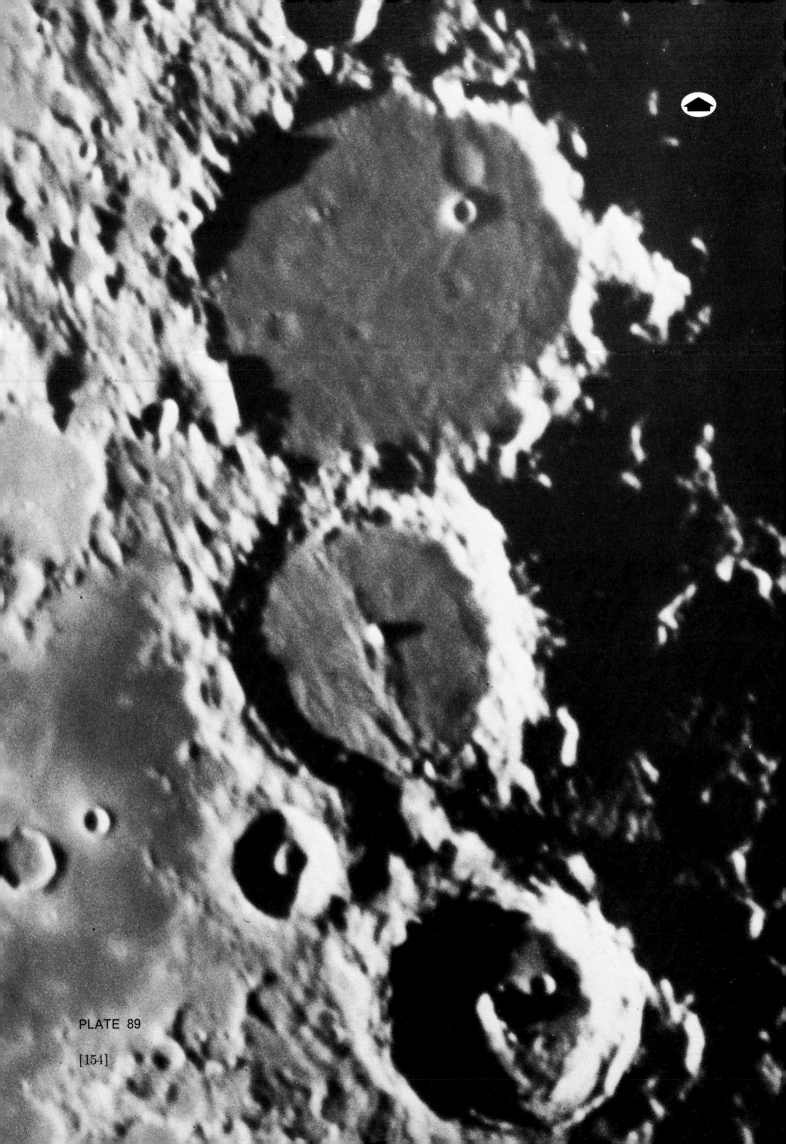

PLATE 89

[154]

PLATE 90

[155]

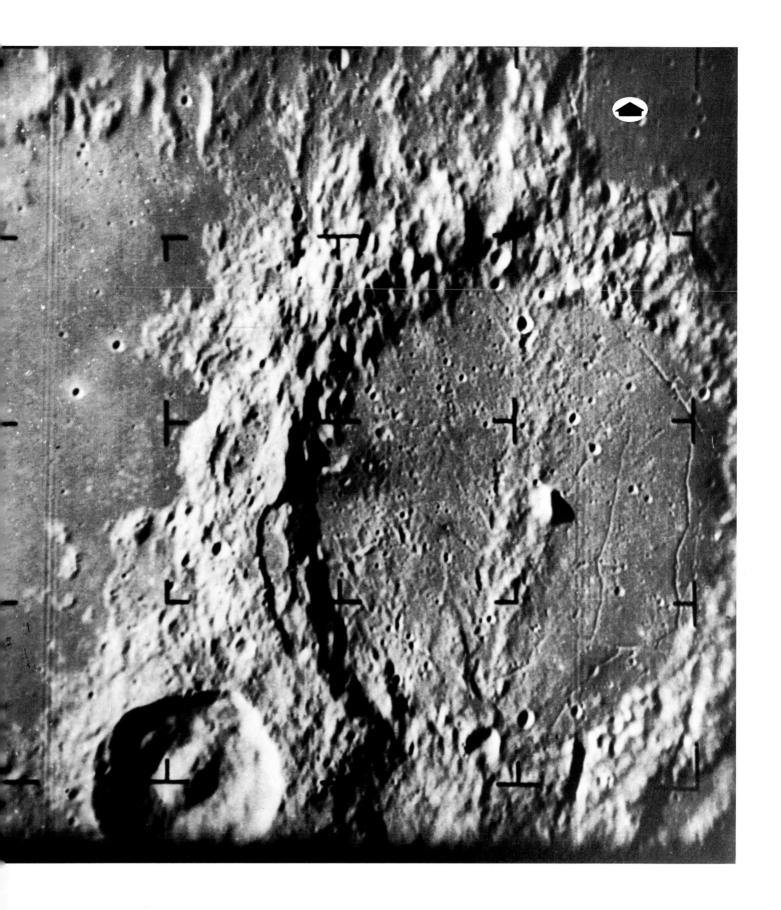

PLATE 91

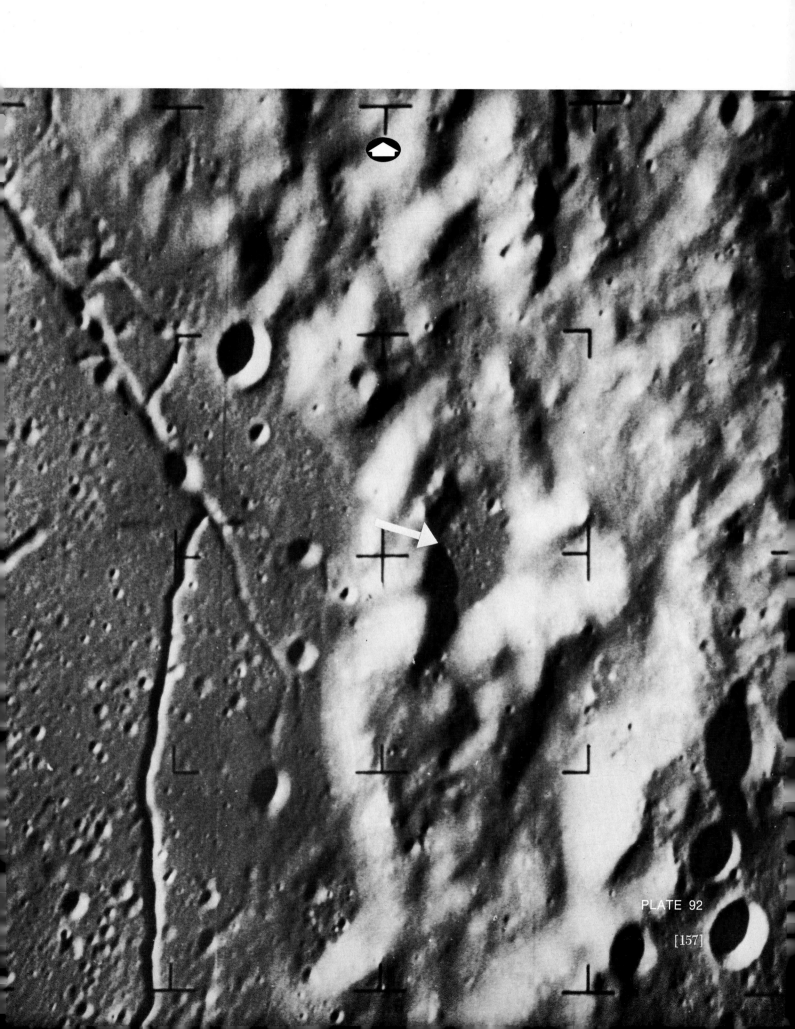

PLATE 92

[157]

PLATE 93

PLATE 94

[159]

PLATE 95

PLATE

[161]

97. The craters Albategnius (lower left) and Hipparchus (center), photographed on October 21, 1968 with the 43-inch reflector at Observatoire du Pic-du-Midi. (Manchester Lunar Program)

98. The craters Ptolemaeus, Albategnius (half visible at the bottom of the field), Hipparchus (above them), and Sinus Medii with Triesnecker. Photograph taken on May 27, 1966 with the 43-inch reflector at Observatoire du Pic-du-Midi. (Manchester Lunar Program)

99. Sunset over the lunar continental landscape south of Mare Nectaris. The Altai scarp (part of an incomplete ring surrounding the mare) is seen to run between the upper left corner and the crater Piccolomini, and the Sun is just setting over the Rheita Valley (lower right corner); the crater Fracastorius (on the southern shores of Mare Nectaris) is at the top. Photograph taken on September 14, 1965 with the 43-inch reflector at Observatoire du Pic-du Midi. (Manchester Lunar Program)

100. Sunset over the Altai scarp, Rheita Valley, and the southern shores of Mare Nectaris (top), photographed on September 14, 1965 with the 43-inch reflector at Observatoire du Pic-du-Midi. (Manchester Lunar Program)

101. The craters Theophilus (center), Cyrillus, and Catharina on the western shores of Mare Nectaris, photographed on August 3, 1965 with the 43-inch reflector at Observatoire du Pic-du-Midi. Note a great number of small craters in the plains of Mare Nectaris, due obviously to secondary impacts at the time Theophilus was formed. (Manchester Lunar Program)

102. Sunset over Mare Nectaris and the southern parts of Mare Tranquillitatis (above), with the craters Fracastorius (bottom) and Theophilus, Cyrillus, and Catharina to the west. Photograph taken on September 15, 1965 with the 43-inch reflector at Observatoire du Pic-du-Midi. (Manchester Lunar Program)

103. Sunset over Theophilus, photographed on July 29, 1964 with the 24-inch refractor at Observatoire du Pic-du-Midi. (Manchester Lunar Program)

104. Advancing sunset over Theophilus. Photograph taken on July 29, 1964 with the 24-inch refractor at Observatoire du Pic-du-Midi, 2½ hours later than the one reproduced on Plate 103. (Manchester Lunar Program)

105. Sunrise photographs of the crater Theophilus (south at top) as photographed from the Earth with the 24-inch refractor at Observatoire du Pic-du-Midi (below) and by Lunar Orbiter 3 on February 22, 1967 from an altitude of 55 km (above); illustrating clearly the pockmark character of lunar craters and the general depression of their floors.

106. Sunrise over the northern part of the lunar hemisphere visible from the Earth dominated by the lunar Alps and the large crater Plato between the northern shores of Mare Imbrium (below) and Mare Frigoris. Photograph taken on March 30, 1966 with the 43-inch reflector at Observatoire du Pic-du-Midi. (Manchester Lunar Program)

107. Sunrise over the north pole of the Moon (above) and the crater Plato between Mare Imbrium and Mare Frigoris, at the western end of the chain of the lunar Alps intersected by the Alpine Valley. Photograph taken on September 17, 1965 with the 43-inch reflector at Observatoire du Pic-du-Midi. (Manchester Lunar Program)

[162]

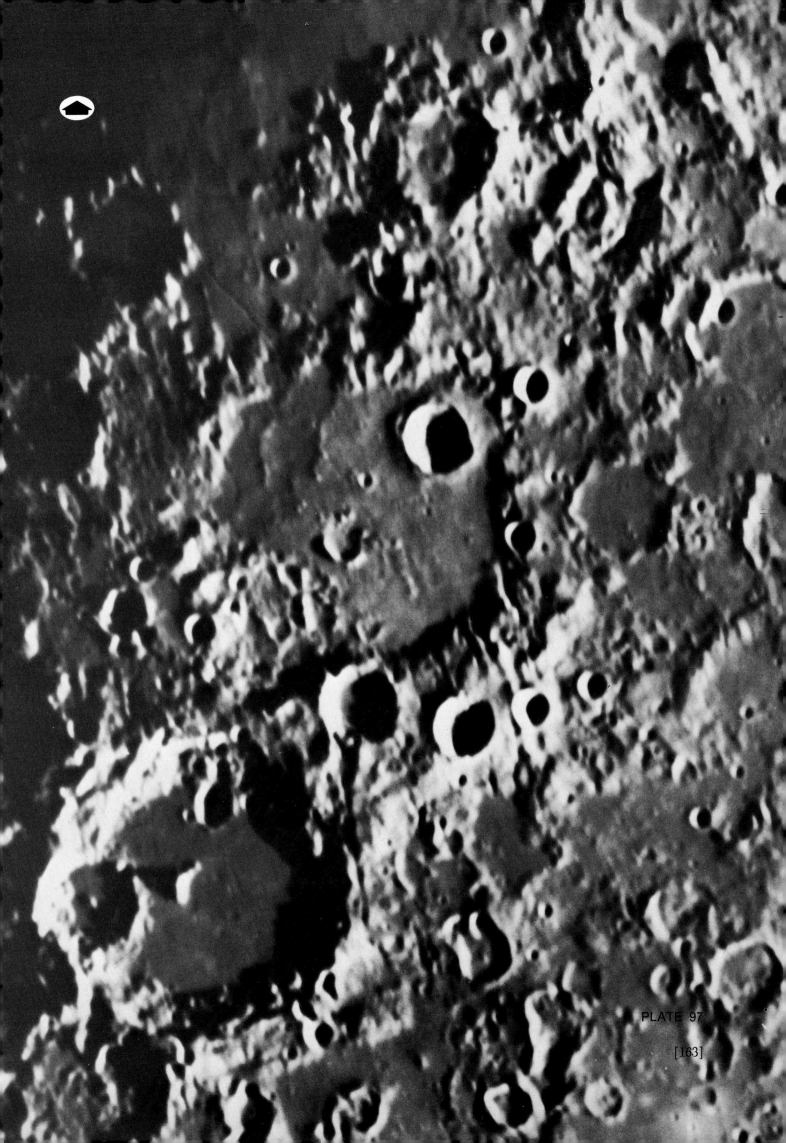

PLATE 97

[163]

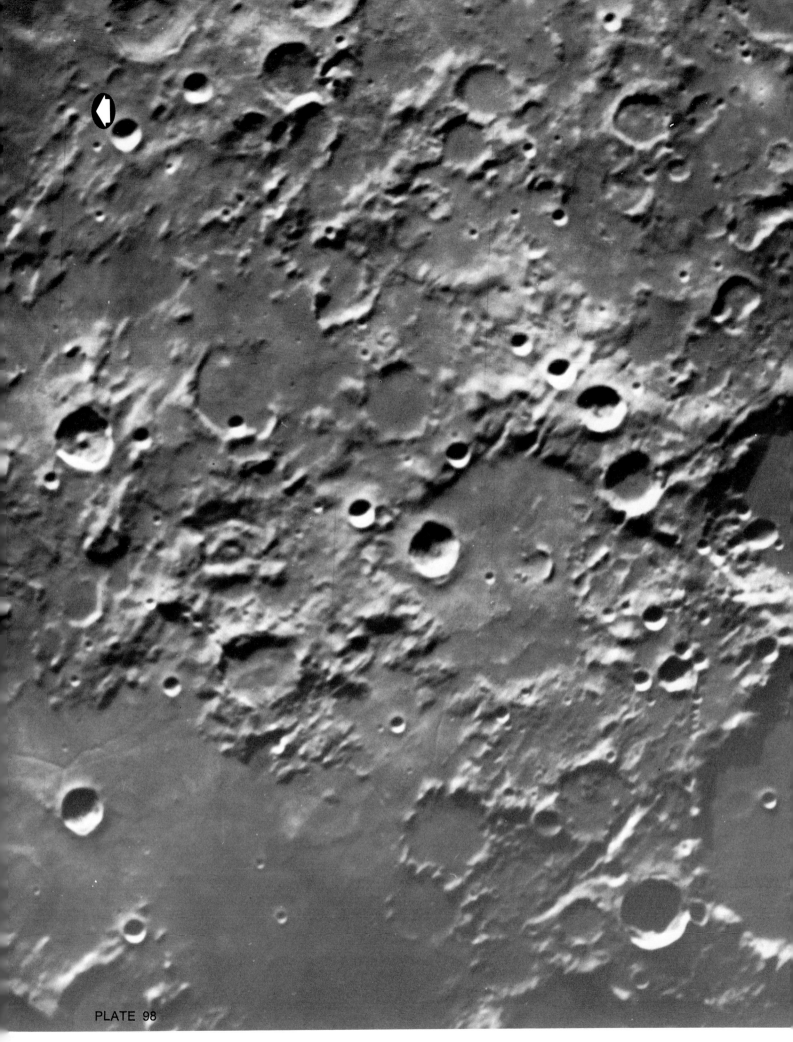

PLATE 98

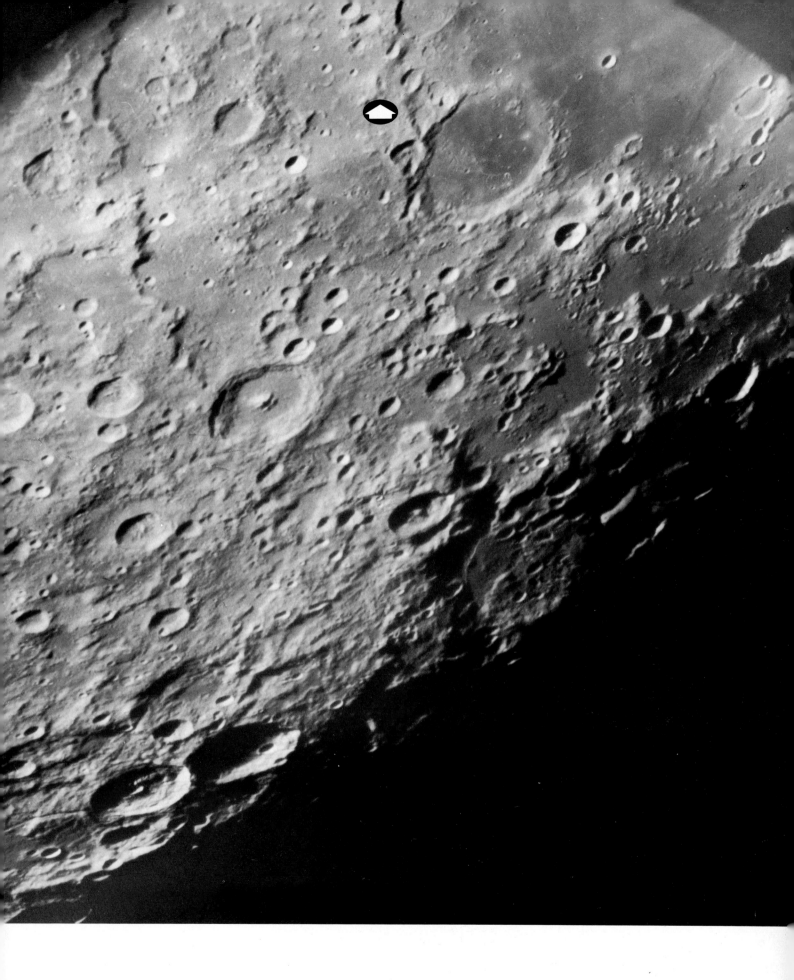

PLATE 99

PLATE 10

[166]

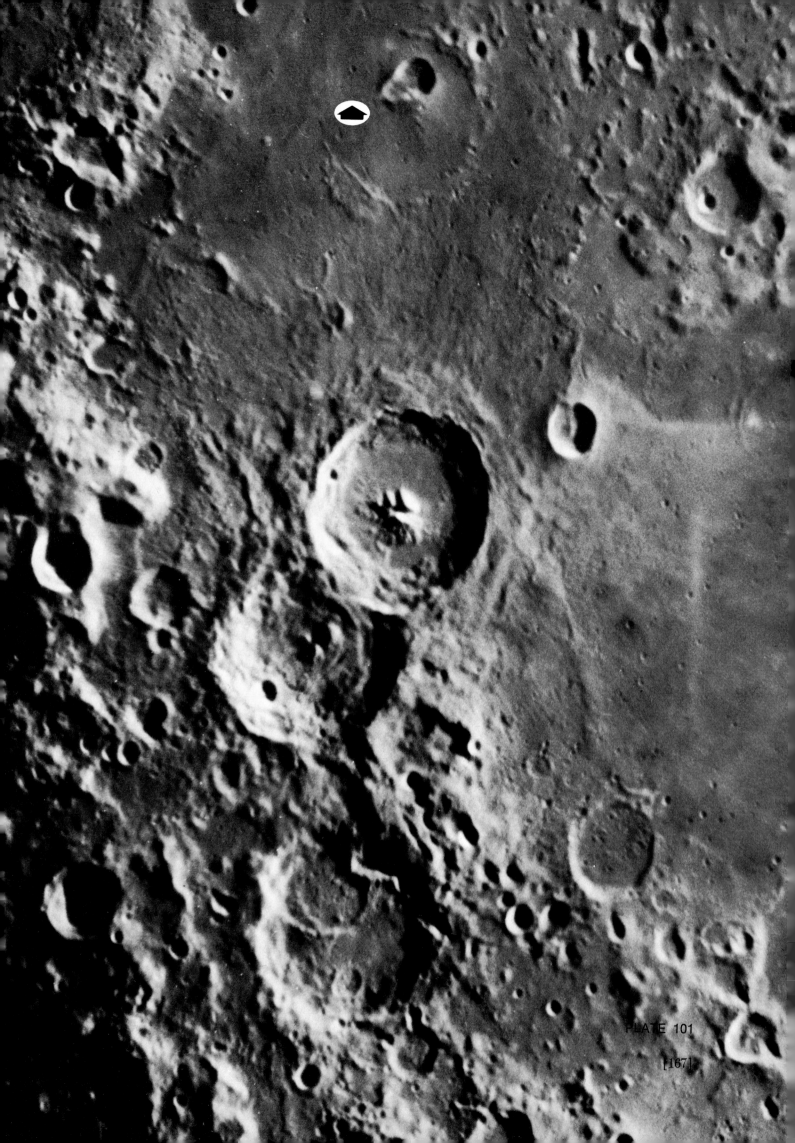

PLATE 101

[167]

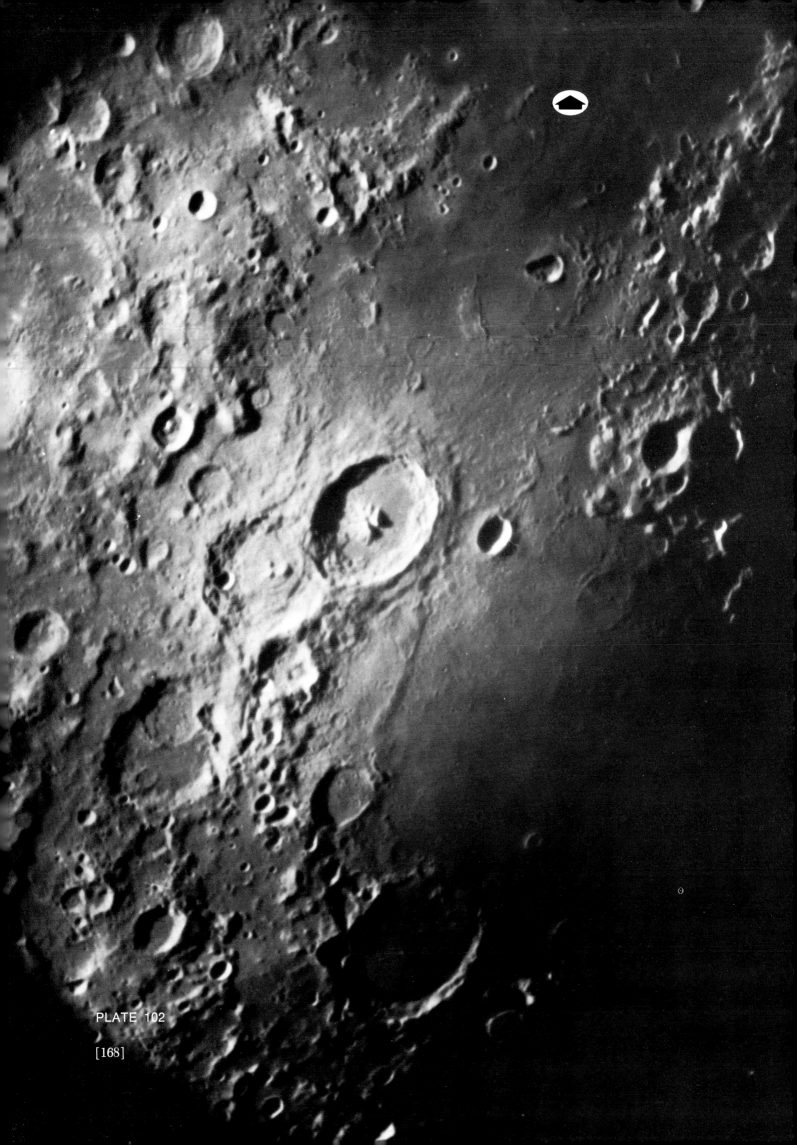

PLATE 102

PLATE 103

[169]

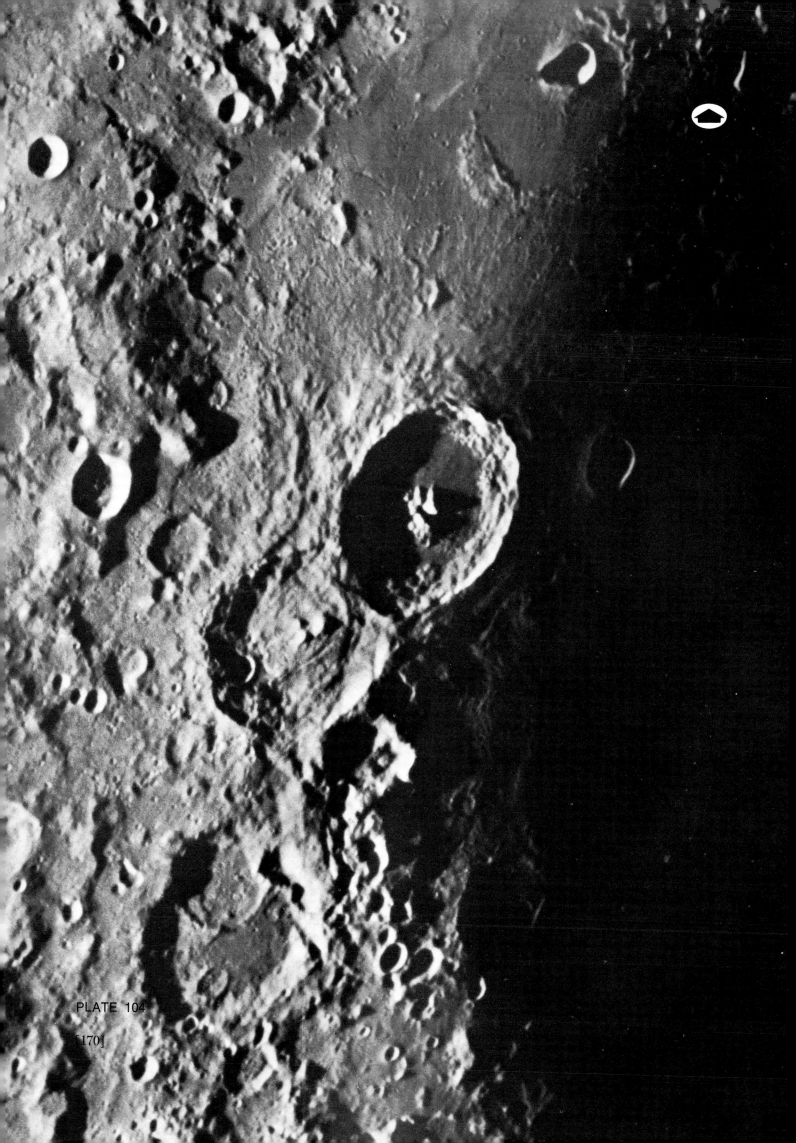

PLATE 104

[170]

TE 105

PLATE 106

[172]

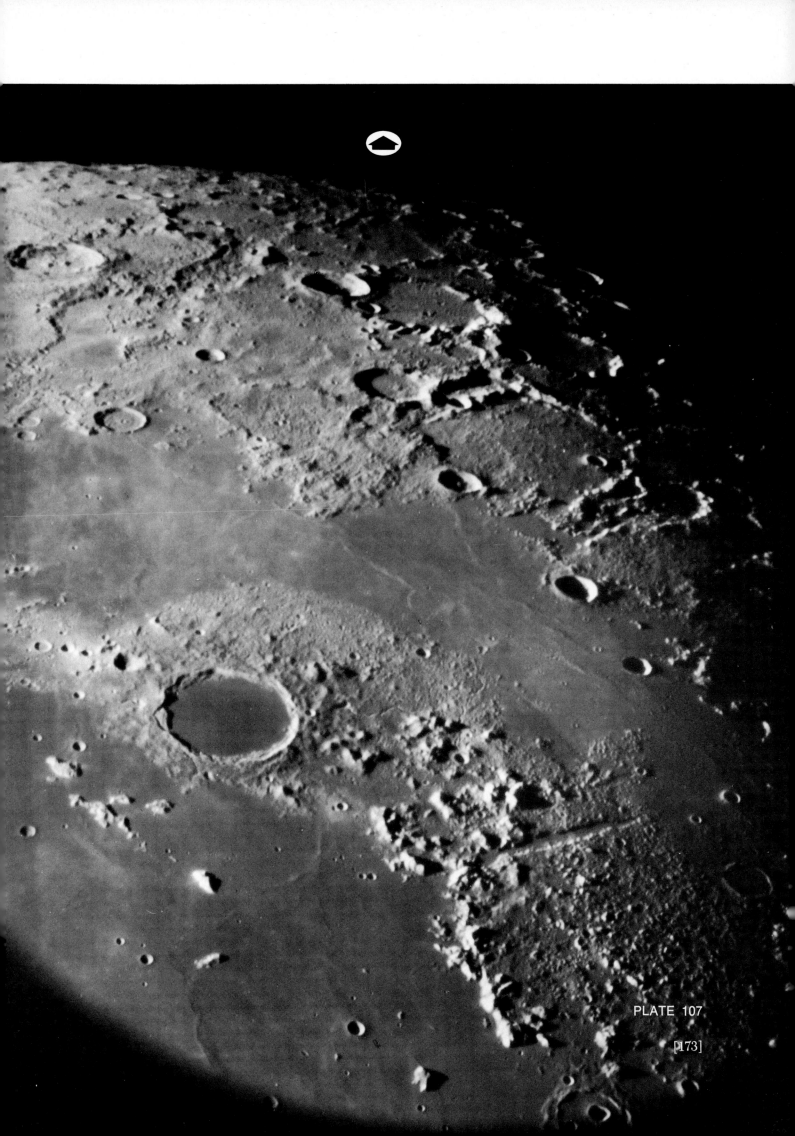

PLATE 107

[173]

108. Advancing sunset over the eastern end of Mare Imbrium and the crater Plato (center). Photograph taken on August 20, 1965 with the 43-inch reflector at Observatoire du Pic-du-Midi (Manchester Lunar Program)

109. A photograph of the crater Plato, taken from an overhead vantage point by the telephoto lens of Lunar Orbiter 4 on May 20, 1967 from an altitude of 2,884 km. (Reproduced by courtesy of NASA-Langley and of the Boeing Company)

110. Sunrise over the lunar crater Eratosthenes at the border between Mare Imbrium to the north (top) and Sinus Aestuum to the south—a typical impact formation on the southern tip of the chain of Apennine Mountains. The outlines of the "ghost crater" Stadius are easily seen at the time of sunrise and so are many of the shallow "secondary" craters associated with the large impact crater Copernicus (just outside the field of view in the lower left-hand corner). Photograph taken on March 30, 1966 with the 43-inch reflector at Observatoire du Pic-du-Midi. (Manchester Lunar Program)

111. Sunrise over a section of the lunar landscape at the crossroads between the southern shores of Mare Imbrium bordered by the Carpathian Mountains (top), Sinus Aestuum to the east (right) and Oceanus Procellarum to the west, dominated by the large impact crater Copernicus, some 90 km across. Photograph taken on September 24, 1966 with the 43-inch reflector at Observatoire du Pic-du-Midi (Manchester Lunar Program) and reproduced to provide a setting for the material of the next ten plates.

112. A photograph of the crater Copernicus (left) and of the adjacent parts of Sinus Aestuum to the east of Copernicus, showing profuse evidence of secondary bombardment by the Copernican ejecta. Such secondary craters— ranging in size from a few kilometers down and, if elongated, oriented radially outward from the primary crater—are strewn out in great numbers across the adjacent mare plains, commencing with a certain distance from the Copernican walls; closer to them all small formations of this type have apparently been filled in by smaller debris at the time of the great primary impact which gave rise to Copernicus. A certain number of small sharp-walled craters visible on the apron of the Copernican ejecta were formed by isolated impacts since the time of the Copernican catastrophe. Photograph taken on May 19, 1967 by the high-resolution lens of Lunar Orbiter 4 from an altitude of 2,680 km above the lunar surface. (Reproduced by courtesy of NASA-Langley and of the Boeing Company)

113. A composite plate of a ground-based photograph of Copernicus (upper right) with two angles of sight marked on it by A and B. View A represents the field covered by the photograph on the upper left, and taken by Lunar Orbiter 2 on November 23, 1966 from a closer proximity to its target (with the camera pointed 17° below the horizon). The photograph below shows a high-resolution view of the interior of Copernicus, taken from the vantage point B on the same day from an altitude of 45.4 km. (Reproduced by courtesy of NASA-Langley and of the Boeing Company)

114. Enlargement of the Lunar Orbiter 2 photograph of the crater Copernicus on the horizon, reproduced in the upper left-hand corner of the preceding plate. The pockmark character of the formation and the depression of its floor below the level of the surrounding landscape is as evident as it was for the crater Theophilus on Plate 105. (Reproduced by courtesy of NASA-Langley and of the Boeing Company)

PLATE 108

[175]

PLATE 109

[176]

PLATE 110

[177]

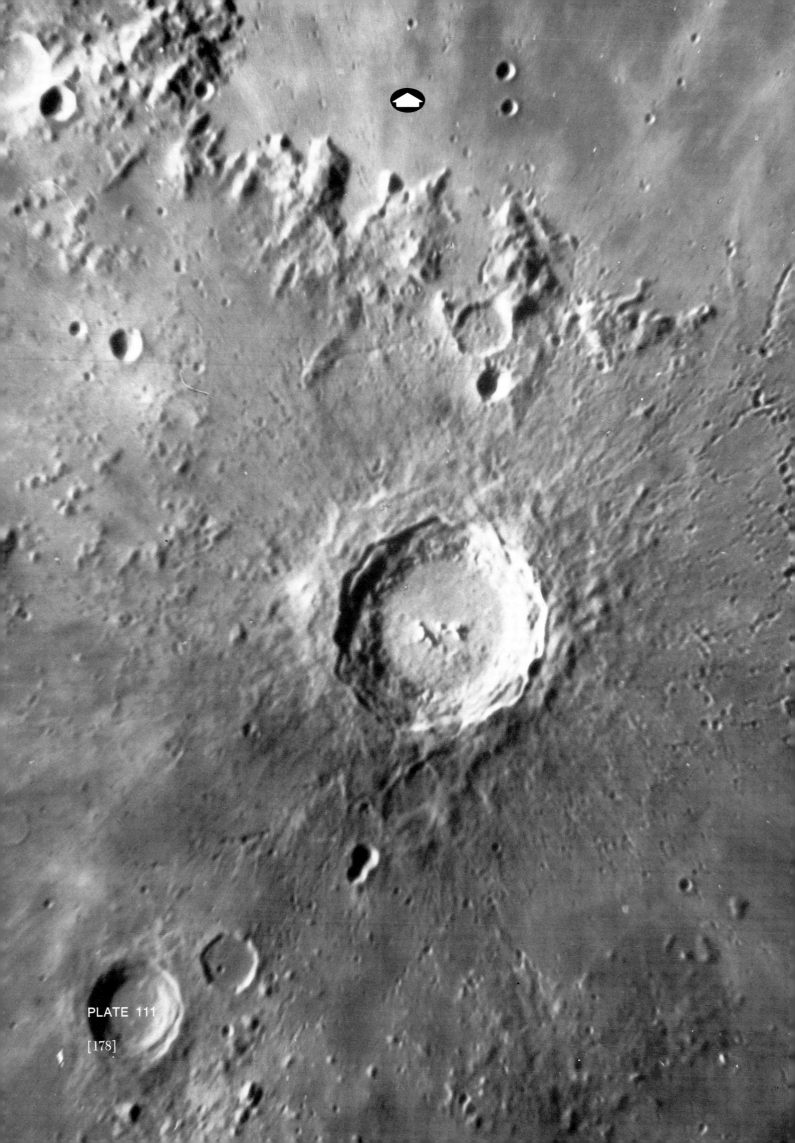

PLATE 111

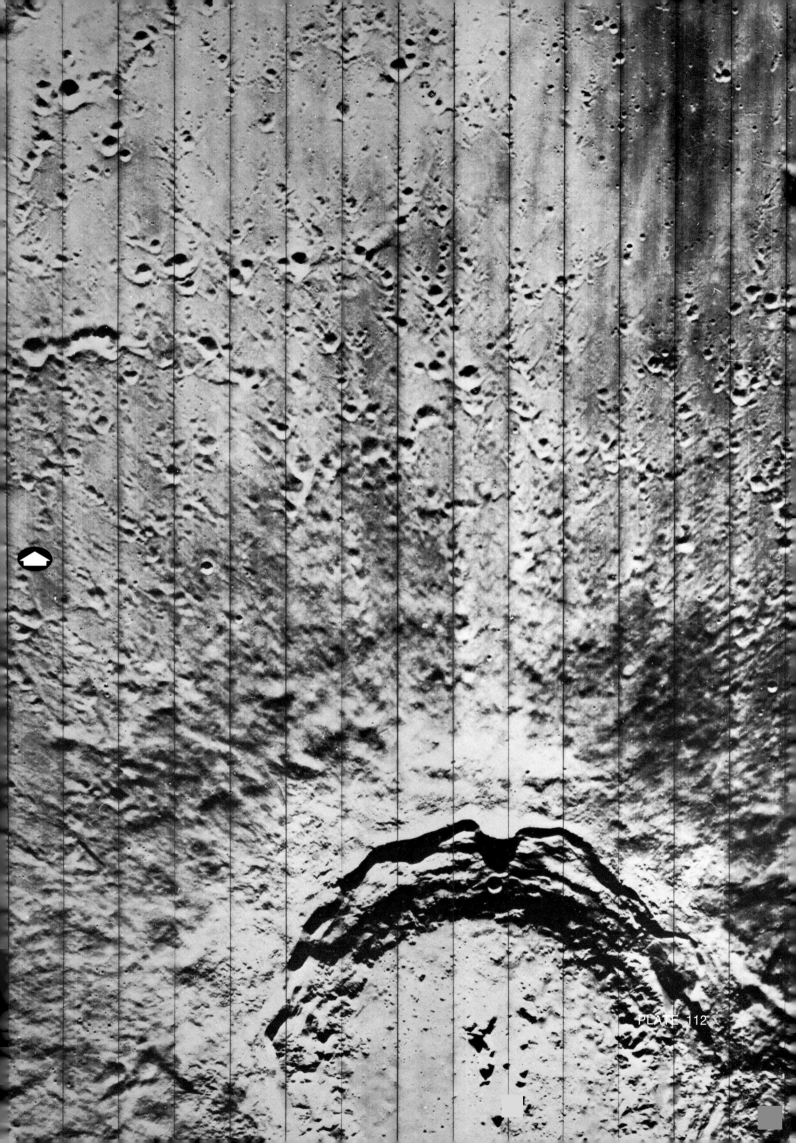

PLATE 112

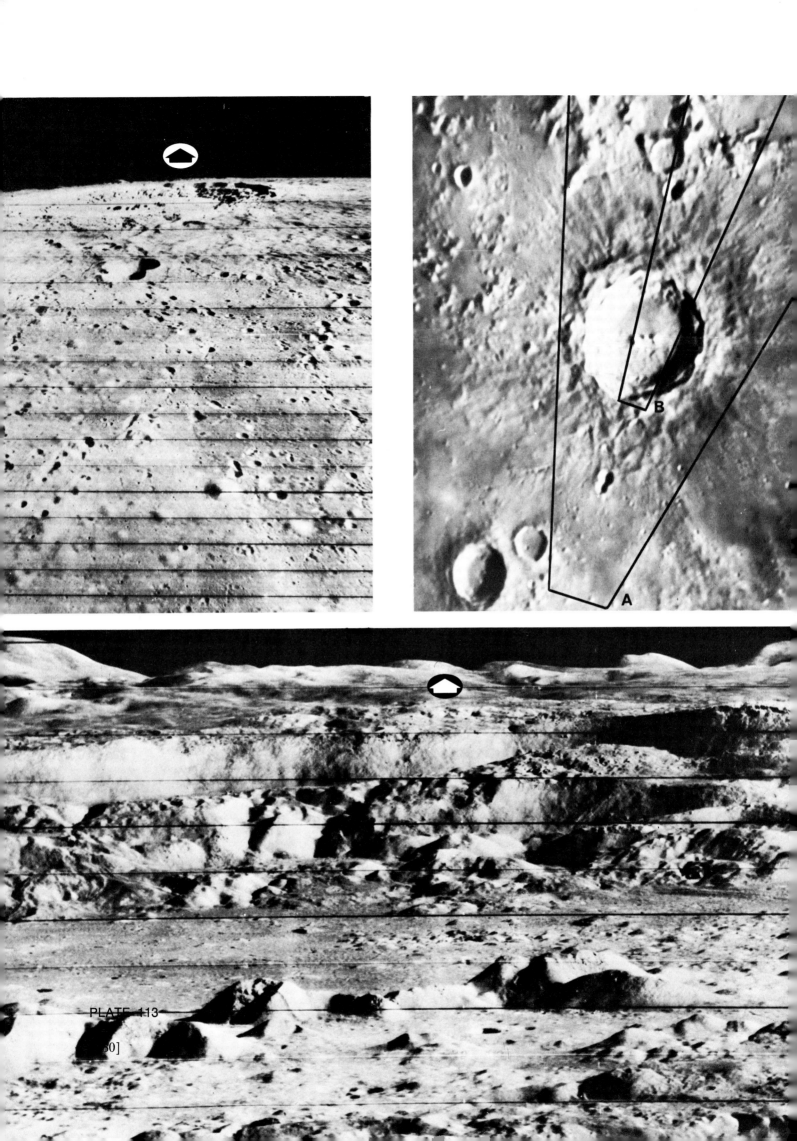

PLATE 113

[90]

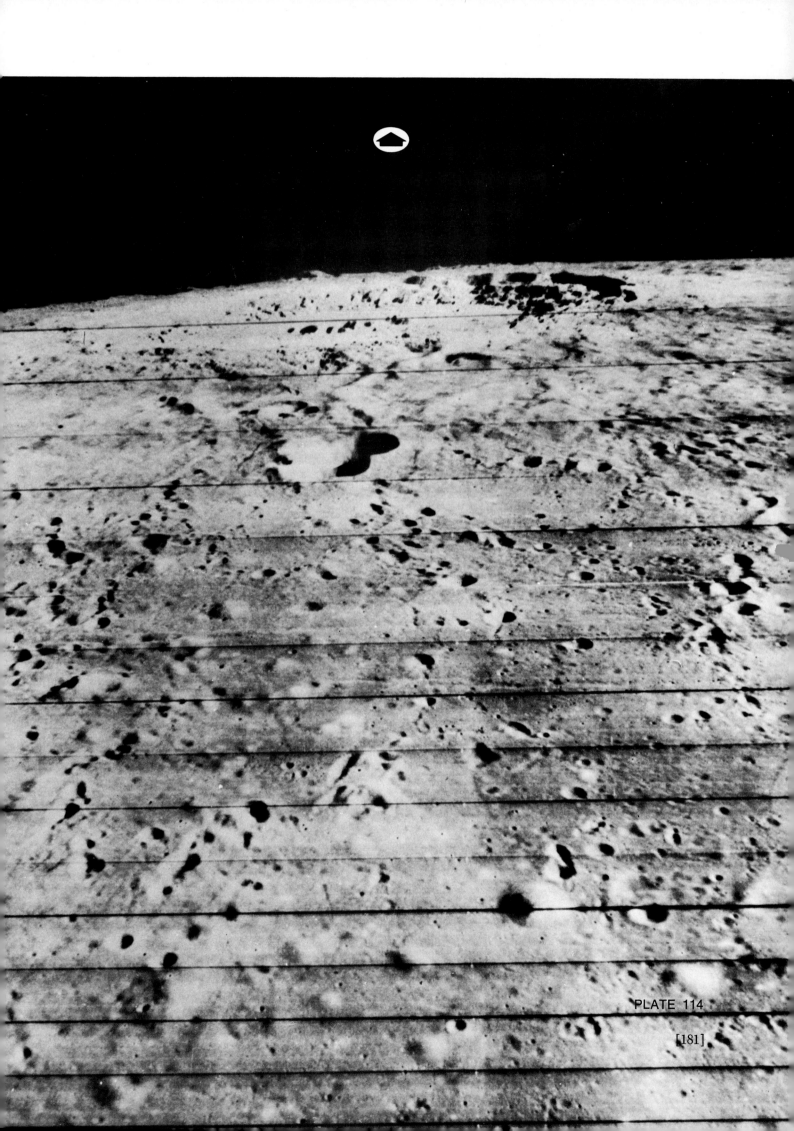

PLATE 114

[181]

115. A view of the interior of the crater Copernicus, as recorded by the tele-photo lens of Lunar Orbiter 2 on November 12, 1966 from an altitude of 46 km above the lunar surface. The depression at the lower end of the photo-graph is the crater Fauth. (Reproduced by courtesy of NASA-Langley and of the Boeing Company)

116. Interior of the crater Copernicus (enlargement of the photograph re-produced in the lower half of Plate 113), as recorded by Lunar Orbiter 2 on November 23, 1966. At the time of the exposure the spacecraft overflew the outward-sloping southern ramparts of the crater at an altitude of 45 km. The group of hills in the forefront constitute the "central mountain" of Copernicus; and the broken nature of the inward-sloping walls of the northern ramparts is seen in the background at a distance close to 90 km. The large hill on the horizon to the left is the 900-m high Gay Lussac of the lunar Carpathians. (Reproduced by courtesy of NASA-Langley and of the Boeing Company)

117. Detail of the central mountain of Copernicus, photographed by the high-resolution lens of Lunar Orbiter 5 on August 16, 1967 from an altitude of 103 km. The ground resolution on this photograph is close to few meters on the lunar surface, and discloses the presence of a great many boulders of this size. (Reproduced by courtesy of NASA-Langley and of the Boeing Company)

118. Northern ramparts of the crater Copernicus, as photographed on August 16,1967 by Lunar Orbiter 5 from an overhead vantage point at an altitude of 104 km. Note the rugged nature of the inward-sloping walls—in contrast with the much gentler incline of those sloping outward. (Reproduced by courtesy of NASA-Langley and of the Boeing Company)

119. A section of the inward-sloping walls of the crater Copernicus, photo-graphed on August 16, 1967 by the high-resolution lens of Lunar Orbiter 5 from an altitude of 104 km. (Reproduced by courtesy of NASA-Langley and of the Boeing Company)

120. Detail of the floor of Copernicus between the central mountain of this crater and the inner rim of its northern ramparts. The white rectangle on the top marks the field of view of the photograph reproduced on Plate 121; the one at the bottom, of Plate 207. (Reproduced by courtesy of NASA-Langley and of the Boeing Company)

121. Detail of the northern part of the floor of Copernicus, shown on Plate 120 and recorded by the telephoto lens of Lunar Orbiter 5 on August 16, 1967 from an altitude of 103 km. The field exhibits several beautiful examples of lunar domes with crater-like orifices on top, from which stones were obviously disgorged. (Reproduced by courtesy of NASA-Langley and of the Boeing Company)

122. The crater Aristoteles, with Mitchell on its side (partly destroyed), Egede (intersected by the lower margin), and Galle near the top, photo-graphed from an overhead position by the telephoto lens of Lunar Orbiter 4 on May 17, 1907 from an altitude of 2,935 km. Note the similarity of the terraced structure of the walls, and the associated family of secondary craters, with those of Copernicus as seen on Plate 112. (Reproduced by courtesy of NASA-Langley and of the Boeing Company)

123. An oblique view of the crater Kepler in Oceanus Procellarum, as recorded by Lunar Orbiter 3 on February 22, 1967 from an altitude of 730 km. (Reproduced by courtesy of NASA-Langley and of the Boeing Company)

[182]

PLATE 115

PLATE 116

[94]

TE 117

PLATE 118

[186]

PLATE 119

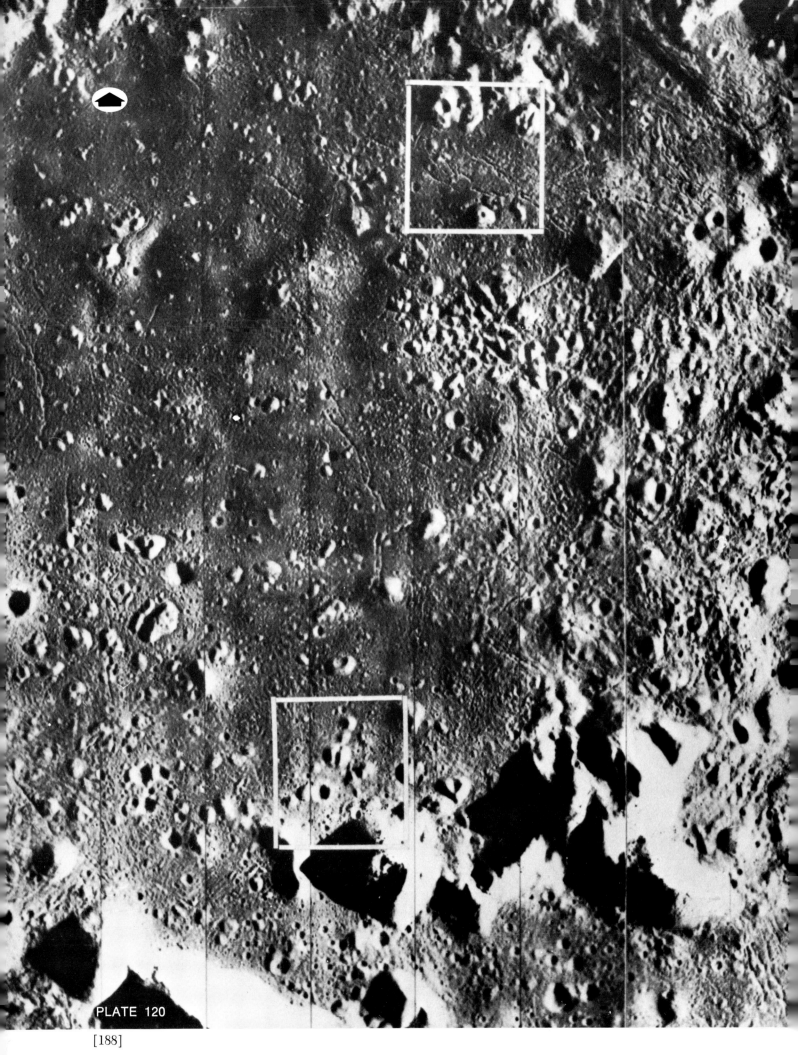

PLATE 120

PLATE 121

PLATE 122

PLATE 123

[191]

124. Sunrise over the twin craters Aristarchus (right) and Herodotus, with the well-known Schröter's Canyon branching out from the "cobra's head" above Herodotus westward. The crater Prinz and its associated system of rilles can be seen in the upper right corner. Photograph taken on May 22, 1967 by the high-resolution lens of Lunar Orbiter 4 from an altitude of 2,668 km. (Reproduced by courtesy of NASA-Langley and of the Boeing Company)

125. A high-resolution photograph of a part of the inward-sloping walls of the crater Aristarchus, taken on August 18, 1967 by Lunar Orbiter 5 from an altitude of 128 km. (Reproduced by courtesy of NASA-Langley and of the Boeing Company)

126. A trace of an avalanche of boulders down the inner slopes of the northwest section of the crater Aristarchus, photographed on August 18, 1967 by Lunar Orbiter 5 from an altitude of 127 km. (Reproduced by courtesy of NASA-Langley and of the Boeing Company)

127. A high-resolution photograph of a part of the floor of Aristarchus, with its central mountain in the lower right corner, taken on August 18, 1967 by Lunar Orbiter 5 from an altitude of 127 km. (Reproduced by courtesy of NASA-Langley and of the Boeing Company)

128. A high-resolution photograph of a section of the rims of Aristarchus sloping outward in the southeast quadrant of the crater, taken on August 18, 1967 by Lunar Orbiter 5 from an altitude of 127 km. (Reproduced by courtesy of NASA-Langley and of the Boeing Company)

129. The characteristic crater Messier in Mare Foecunditatis, as photographed on August 14, 1967 by Lunar Orbiter 5 from an altitude of 98 km. (Reproduced by courtesy of NASA-Langley and of the Boeing Company)

130. A photograph of the crater Goclenius on the western shores of Mare Foecunditatis, photographed at an oblique angle (looking southward) from Apollo 8 spacecraft on December 24, 1968. The floor of Goclenius is seen to be crossed by prominent rilles. The fact that the largest of them cuts across both the northern rim and the central peak of Goclenius indicates that this rille was formed after the crater came into being and represents, therefore, a surface manifestation of deeper structure. Other larger craters in the background include (left to right) Colombo A, Magelhaens and Magelhaens A. (Reproduced by courtesy of NASA)

131. The crater Lichtenberg with its "ghost" (possessing a "central peak") in the northwestern region of the Oceanus Procellarum. The crater Briggs (flanked by Briggs B and C on the left and right) near the lower end of the frame, with a triangle of Narimann and Narimann B and G near the top. Photograph taken on May 23, 1967 by the high-resolution lens of Lunar Orbiter 4 from an altitude of 2,868 km. (Reproduced by courtesy of NASA-Langley and of the Boeing Company)

132. The "ghost ring" around the crater Flamsteed in Oceanus Procellarum; a typical example of formations which look like submerged rims of large craters. Photograph taken on May 21, 1967 with the high-resolution lens of Lunar Orbiter 4 from an altitude of 2,717 km. Small white square marks the area shown at high resolution on Plate 133. (Reproduced by courtesy of NASA-Langley and of the Boeing Company)

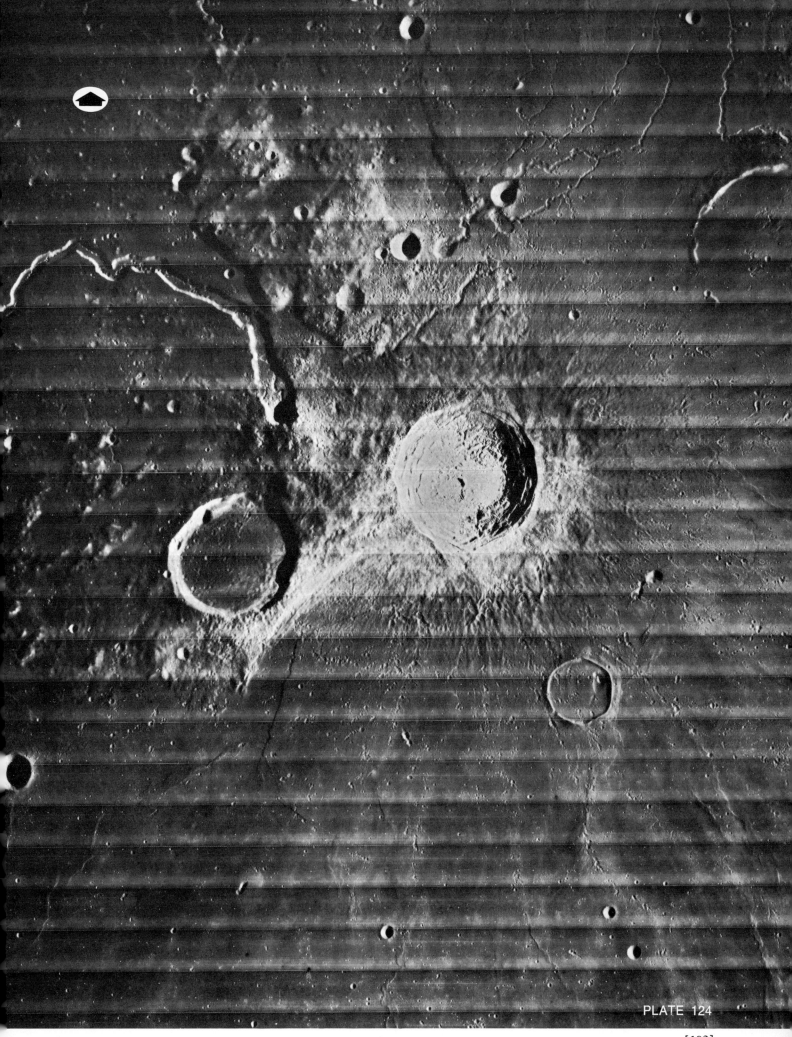

PLATE 124

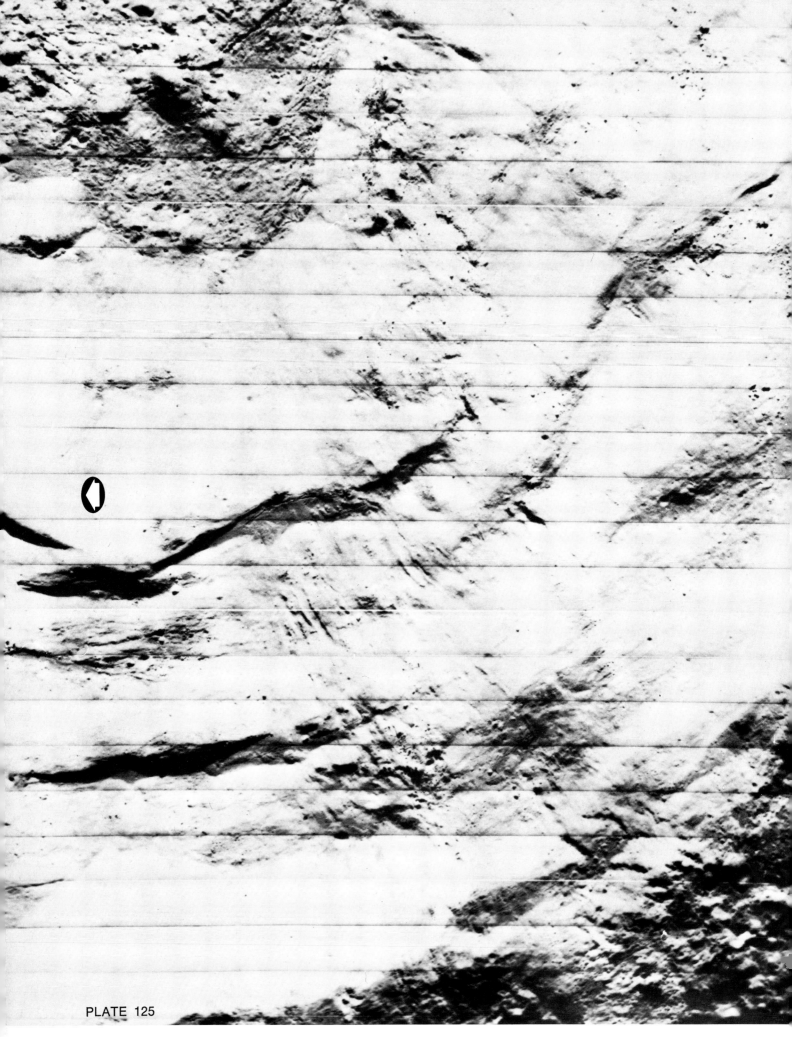

PLATE 125

PLATE 126

[195]

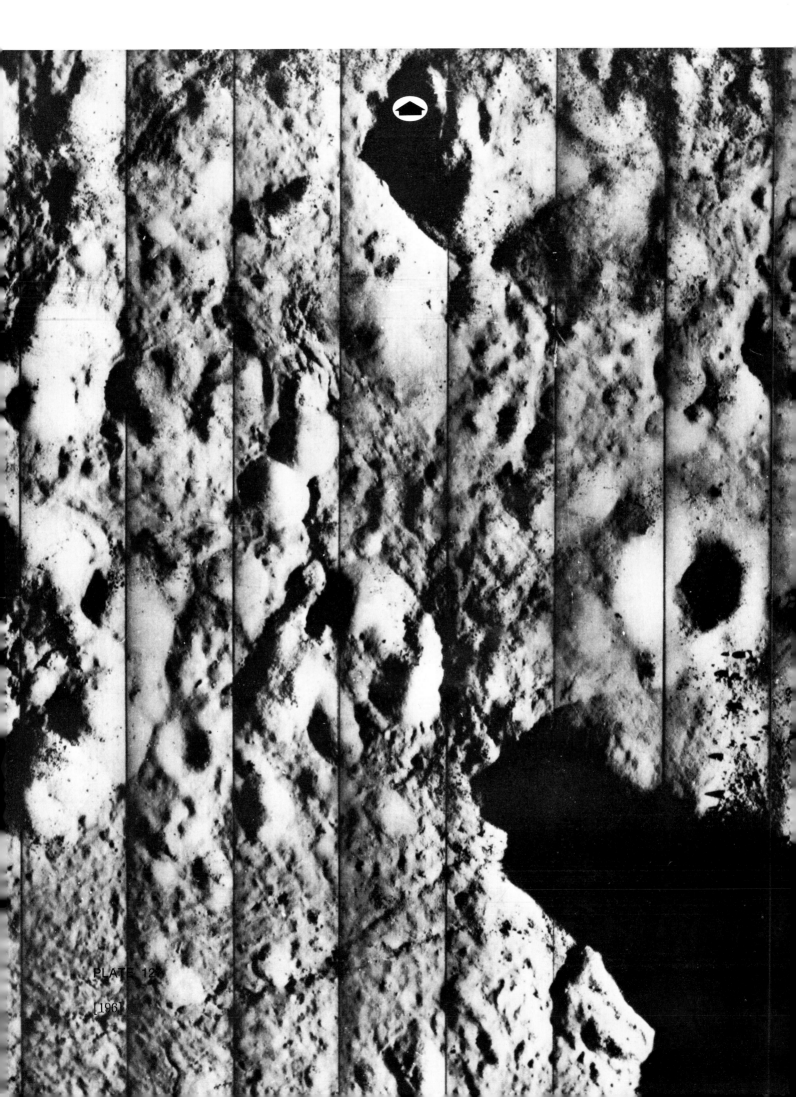

PLATE 12

[196]

28

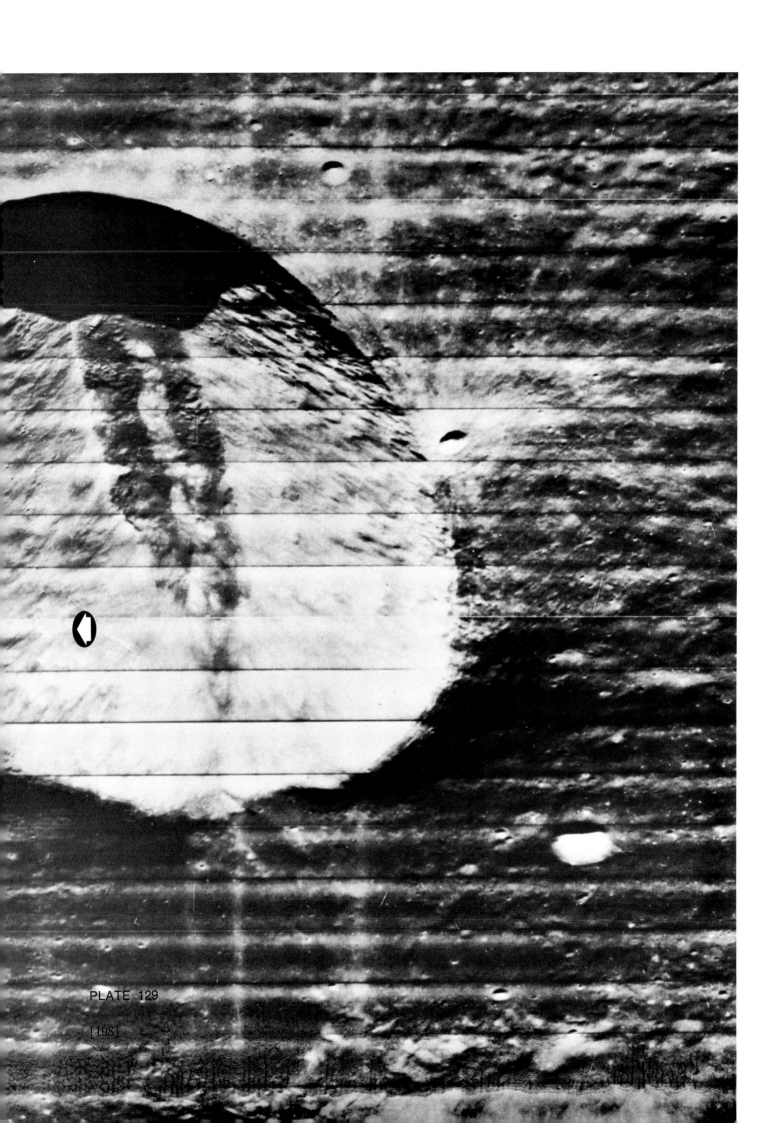

PLATE 129

[198]

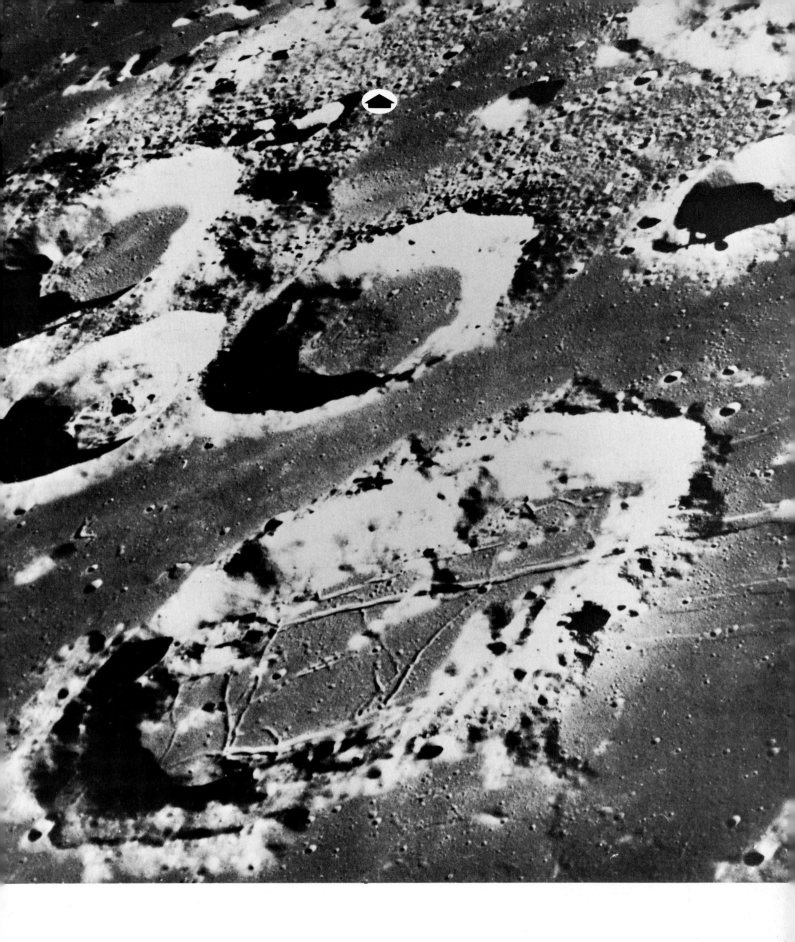

PLATE 130

PLATE 131

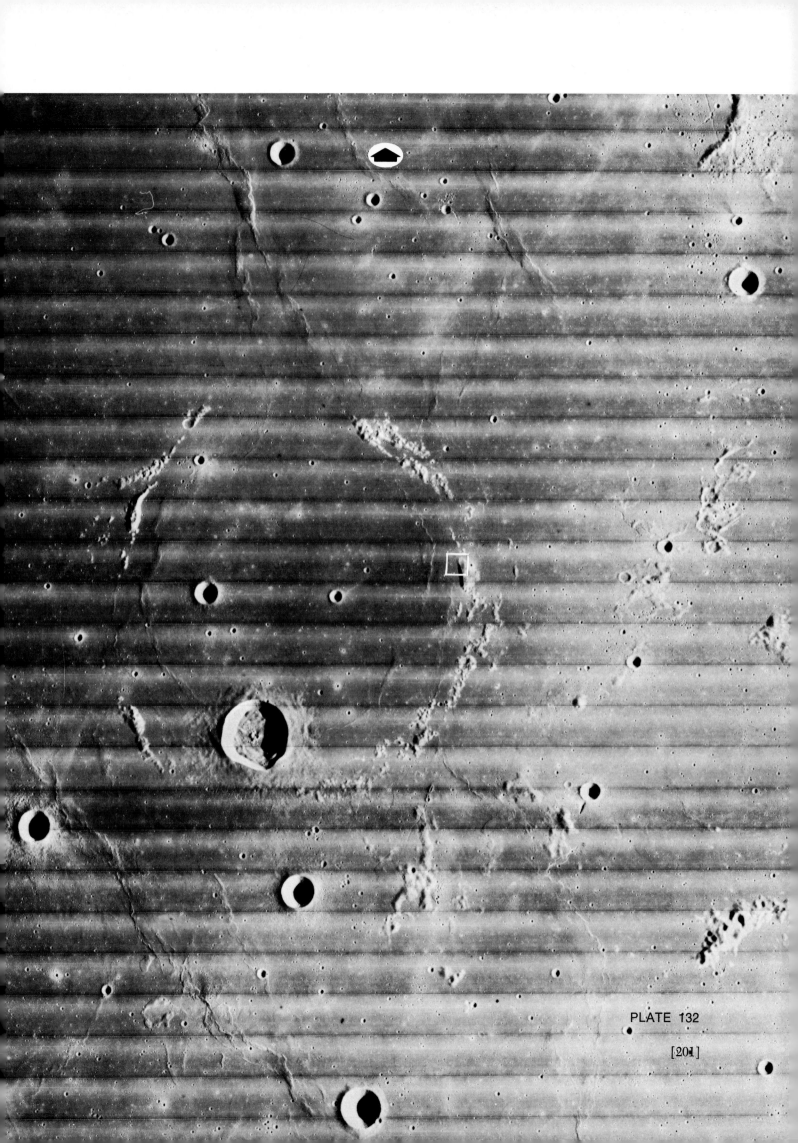

PLATE 132

[211]

133. Western slopes of one of the cliffs constituting the Flamsteed "ghost ring" shown on the preceding plate (the position of the respective cliff is marked by a white arrow), as photographed on February 22, 1967 by the high-resolution lens of Lunar Orbiter 3 from an altitude of 42 km. (Reproduced by courtesy of NASA-Langley and of the Boeing Company)

134. Sunrise over the Moon's far side, as photographed by the wide-angle lens of Lunar Orbiter 5 on August 11, 1967 from an altitude of 1,189 km. Unlike the lunar hemisphere facing us, most of the craters on the Moon's far side lack as yet any proper names. The large crater on the terminator line near the upper end of the frame is Oppenheimer; and the large two craters near the terminator are Leibnitz and von Kármán, respectively. (Reproduced by courtesy of NASA-Langley and of the Boeing Company)

135. Evening crescent of the Moon's far side featuring Mare Moscoviense (below) and above it two large craters are yet unnamed, at selenographic coordinates $\lambda = 154°5$ E, $\beta = 45°0$ N and $\lambda = 165°0$ E, $\beta = 51°0$ N, respectively. The crater marked with a white arrow above Mare Moscoviense is Kurchatov. Photograph taken on August 15, 1967 by the wide-angle lens of Lunar Orbiter 5 from an altitude of 1,234 km. (Reproduced by courtesy of NASA-Langley and of the Boeing Company)

136. Sunset over the Mare Moscoviense, the largest (and, in fact, the only) mare on the Moon's far side west of Mare Orientale on the opposite limb of the Moon. Photographed on August 15, 1967 with the wide-angle lens of Lunar Orbiter 5 from an altitude of 1,234 km. (Reproduced by courtesy of NASA-Langley and of the Boeing Company)

137. A typical landscape of the Moon's far side, replete with innumerable craters. Most of the smaller craters shown on this plate have as yet received no proper names; the large crater near the bottom of the field is Gagarin; the large formation in the upper left is Mendeleev. Photograph taken on August 25, 1966 by the wide-angle lens of Orbiter 1 from an altitude of 1,454 km. (Reproduced by courtesy of NASA-Langley and of the Boeing Company)

138. Margin of the Moon's visible hemisphere in the east featuring Mare Marginis and Mare Smythii in the upper left corner (with the crater Napier in between) which can still be seen from the Earth at favorable librations. However, the large craters Pasteur (near the center) or Sklodovska-Curie (to the west of it, with a central peak) are no longer visible from the Earth. This photograph was taken on November 25, 1966 by Lunar Orbiter 2 from an overhead vantage point at an altitude of 1,517 km. (Reproduced by courtesy of NASA-Langley and of the Boeing Company)

139. Regions of the Moon's far side in the direction of the south pole between 105°–145° of eastern longitude. The crater Pasteur (see the preceding plate) is bisected by the margin of the frame to the left; and the great crater Tsiolkovsky (with the marelike floor and central mountain) is seen near the center of the field. The ridge of "Soviet Mountains" between Pasteur and Tsiolkovsky, whose discovery was claimed by the Russians from the first photographs of the Moon's far side in October 1959, has proved to be an interlocking chain of crater walls. Photograph taken on February 19, 1967 by the wide-angle lens of Lunar Orbiter 3 from an altitude of 1,461 km. (Reproduced by courtesy of NASA-Langley and of the Boeing Company)

140. Far side of the Moon north of the crater Tsiolkovsky, photographed on August 26, 1966 by the wide-angle lens of Lunar Orbiter 1 from an altitude of 1,321 km. The large crater in the upper right corner of the field is Mendeleev. (Reproduced by courtesy of NASA-Langley and of the Boeing Company)

PLATE 133

PLATE 134

[204]

PLATE 135

[205]

PLATE 136

[206]

PLATE 137

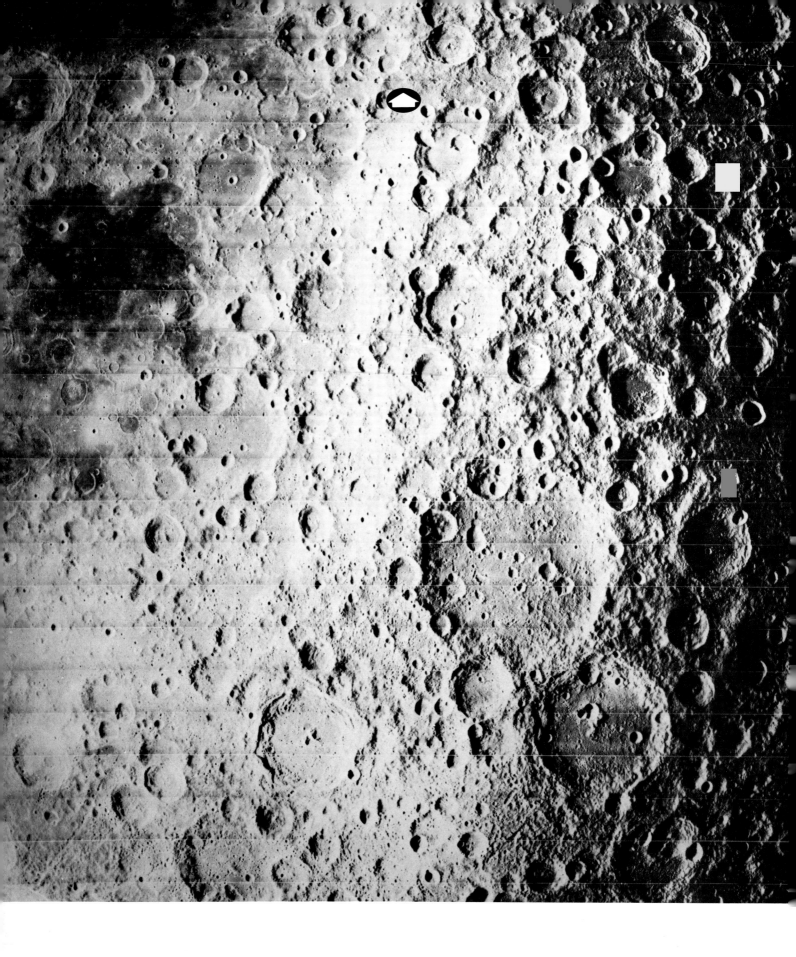

PLATE 138

[208]

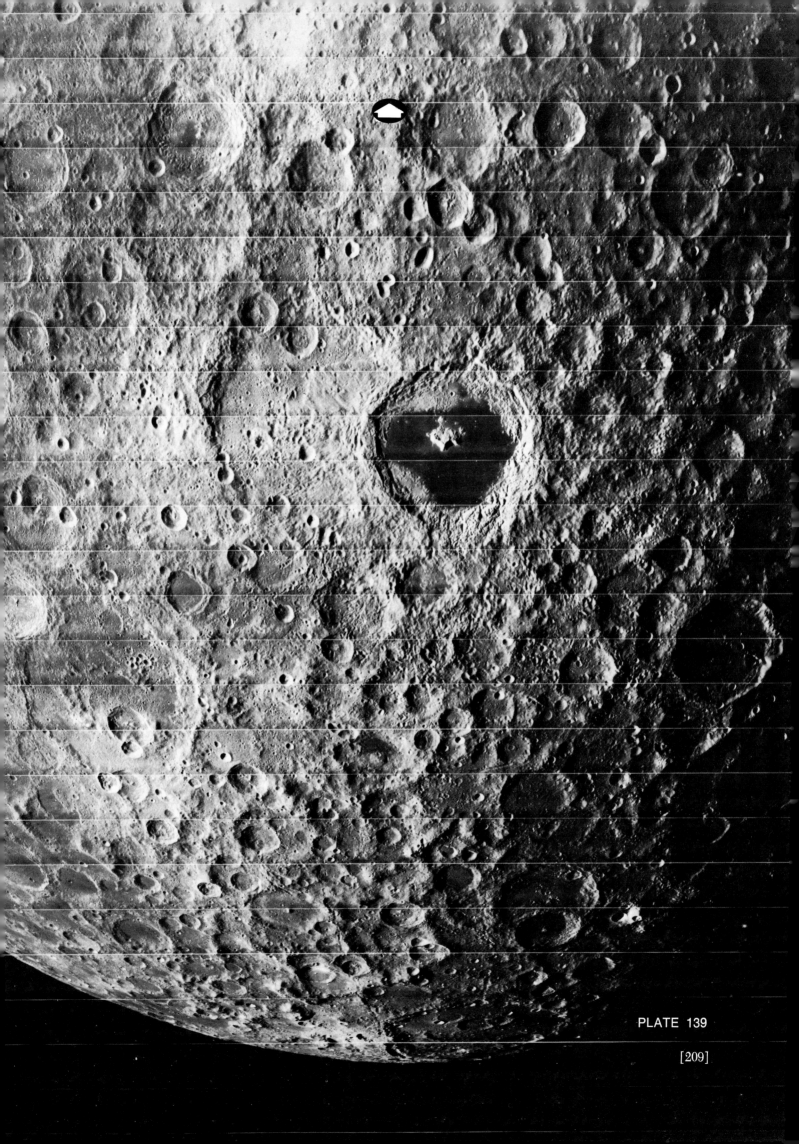

PLATE 139

PLATE 140

141. A high-resolution photograph of a part of the crater Tsiolkovsky with its central mountain, taken on February 19, 1967 by Lunar Orbiter 3 from an altitude of 1319 km. (Reproduced by courtesy of NASA-Langley and of the Boeing Company)

142. Apollo 8 oblique view of the lunar far side to the southeast, with the crater Tsiolkovsky in the center of the field. (Reproduced by courtesy of NASA-Houston)

143. A westward oblique view of the Moon's far side, photographed on August 23, 1966 by Lunar Orbiter 1 from an altitude of 1,170 km above the lunar surface, behind the limb of which we see the Earth in the distance. The large crater seen near the upper part of the field, about 80 km in diameter, is Meitner; that extending farther to the horizon ("below" Earth), 220 km across, is Pasteur. (Reproduced by courtesy of NASA-Langley and of the Boeing Company)

144. Apollo 8 oblique northward view of the far side of the Moon, showing (to the left) the dark-floored crater Lomonosov; and beyond it, the bright-ray crater Giordano Bruno ($\lambda = 102°$ E, $\beta = 38°$ N), whose rays to the southwest cross the east limb of the visible hemisphere of our satellite, and are discernible from the Earth. (Reproduced by courtesy of NASA-Houston)

145. Mare Ingenii, a large crater-like enclosure in the southern part of the Moon's far side ($\lambda = 165°$ E, $\beta = 34°$ S), east of the crater Jules Verne dark floor, to the left of the mare on the print $\lambda = 147°$ E, $\beta = 35°$ S. Photograph taken on November 20, 1966 by Lunar Orbiter 2 from an altitude of 1,466 km. (Reproduced by courtesy of NASA-Langley and of the Boeing Company)

146. A photograph of the large (double-ringed) crater Korolev on the far side of the Moon ($\lambda = 158°$ W, $\beta = 5°$ S), with the crater Doppler immediately below it, and Galois to the right (east) of Doppler, taken on August 20, 1966 by Lunar Orbiter 1 from an altitude of 1,450 km. The black square marks the limits of the field shown on Plate 147. (Reproduced by courtesy of NASA-Langley and of the Boeing Company)

147. Apollo 8 photograph of the floor of the northwest part of the crater Korolev shown on the preceding plate (with position marked by a black square); but whereas the Sun stood relatively high on the latter, the landscape photographed here was illuminated by the Sun standing only 9° above the horizon. (Reproduced by courtesy of NASA-Houston)

148. A photograph of sunrise over the north pole of the Moon, taken by the wide-angle lens of Lunar Orbiter 4 on May 11, 1967 from an altitude of 3,503 km. The large double-walled basin dominating this part of the lunar landscape is Schrödinger; the one to the north of it (in the direction of the narrow gorge emanating from Schrödinger) is Lyot; while Humboldt is near the top. To the left of Schrödinger on the terminator is Amundsen; and below Schrödinger, Drygalski (the lunar south pole is situated between Amundsen and Drygalski). (Reproduced by courtesy of NASA-Langley and of the Boeing Company)

149. An enlargement of a photograph of the crater Schrödinger with its northward gorge, taken by Lunar Orbiter 4 on May 18, 1967 from an altitude of 3,533 km above the lunar surface. (Reproduced by courtesy of NASA-Langley and of the Boeing Company)

150. Floor structure of old craters—Hevelius (see also Plate 52)—as photographed on February 23, 1967 by Lunar Orbiter 3 from an altitude of 63 km above the lunar surface. (Reproduced by courtesy of NASA-Langley and of the Boeing Company)

PLATE 1

[212]

PLATE 142

PLATE 143

[214]

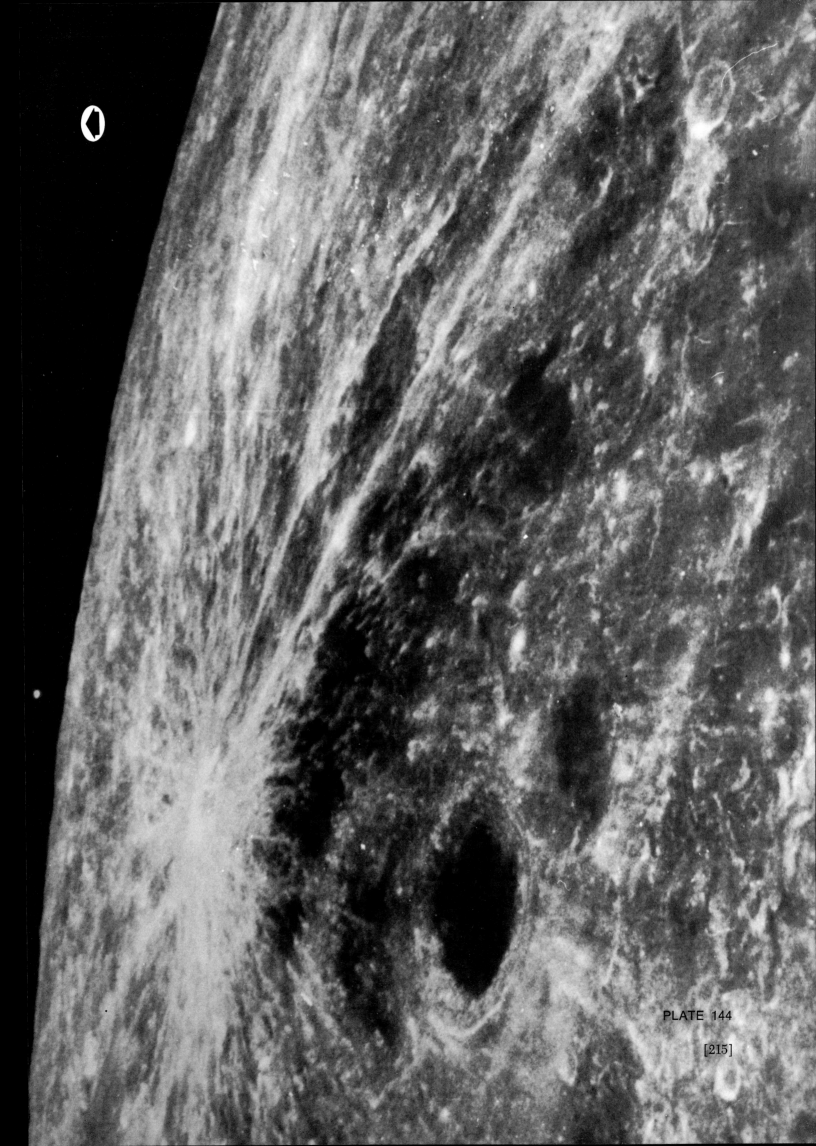

PLATE 144

[215]

PLATE 145

PLATE 146

[217]

PLATE 147

PLATE 148

[219]

PLATE 149

[220]

PLATE 50

1221

151. Another example of broken floor structure; this time of the crater Viliev 24 km in size on the far side of the Moon ($\lambda = 145°8$ E, $\beta = 4°5$ S; between Mendeleev and Gagarin). Photograph taken on August 24, 1966 by the high-resolution lens of Lunar Orbiter 1 from an altitude of 1,379 km above the lunar surface. (Reproduced by courtesy of NASA-Langley and of the Boeing Company)

152. A third example of broken floor structure of lunar craters—Humboldt near the southeastern limb of the visible lunar hemisphere ($\lambda = 81°$ E, $\beta = 26°$ S)—as photographed on May 12, 1967 by the high-resolution lens of Lunar Orbiter 4 from an altitude of 2,745 km. (Reproduced by courtesy of NASA-Langley and of the Boeing Company)

153. The crater Pitatus on the southern shores of Mare Nubium showing a beautiful example of a crater floor broken by rilles which parallel the hexagonal crater walls even more closely than in the case of Alphonsus (Plate 88 or 89). Photograph taken on May 19, 1967 by the high resolution lens of Lunar Orbiter 4 from an altitude of 2,939 km. (Reproduced by courtesy of NASA-Langley and of the Boeing Company)

154. Sunrise over the western shores of the crater-studded Sinus Medii (very nearly the center of the apparent lunar disk visible from the Earth), west of the crater Bruce (just off the field to the right), as photographed from Apollo 10 in May 1969. (Reproduced by courtesy of NASA-Houston)

155. The shallow Hypatia rilles near the southern shores of Mare Tranquillitatis, as televised on February 20, 1965 by Ranger 8, from an altitude of 379 km (for other views of these rilles from different vantage points, see also Plates 156 and 158). The crater bisected by the margin in the upper left corner is Sabine. (Reproduced by courtesy of NASA and of JPL)

156. Southern shores of Mare Tranquillitatis, north of the shallow Hypatia rille (bottom of the field), with the crater Moltke (6.5 km across). Apollo 11 mission landed in the area of this frame (at a spot characterized by the lunar coordinates $\lambda = 23°5$ E, $\beta = 0°8$ N), taken from Apollo 10 in May 1969. Apart from ubiquitous circular craters of all sizes checkering most of the mare ground, note the presence of a number of shallow elongated depressions, 1–2 km in size, discovered previously in this region by Ranger 8. (Reproduced by courtesy of NASA-Houston)

157. A close-up of one of the elongated depressions in Mare Tranquillitatis, as photographed by Apollo 10 in May 1969. Whether such formations are due to grazing impacts or subsidence of the regolith remains as yet uncertain. (Reproduced by courtesy of NASA-Houston)

158. The Apollo 11 astronauts' view of the Moon in the final stage of their descent, 3½ min before their landing on its surface at 6 hours 14 min Universal Time on July 20, 1969. Their landing site is on the sunrise line just north (to the right) of the small crater Moltke, north of the Hypatia rille which we saw on Plate 154 from an overhead position, and which here we see on the horizon. The relatively large crater in the lower right corner of the field is Maskelyne. (Reproduced by courtesy of NASA-Houston)

159. The crater Cauchy (12 km across) and its "hyperbolae" on the eastern shores of Mare Tranquillitatis—the one to the north (above Cauchy) is a shallow rille; while the one to the south is a fault—a cleft analogous to the Straight Wall in Mare Nubium (see Plate 58 or 59)—to the south of which we see two typical domes. Photograph taken on November 23, 1964 with the 43-inch reflector of the Observatoire du Pic-du-Midi. (Manchester Lunar Program)

[222]

PLATE 151

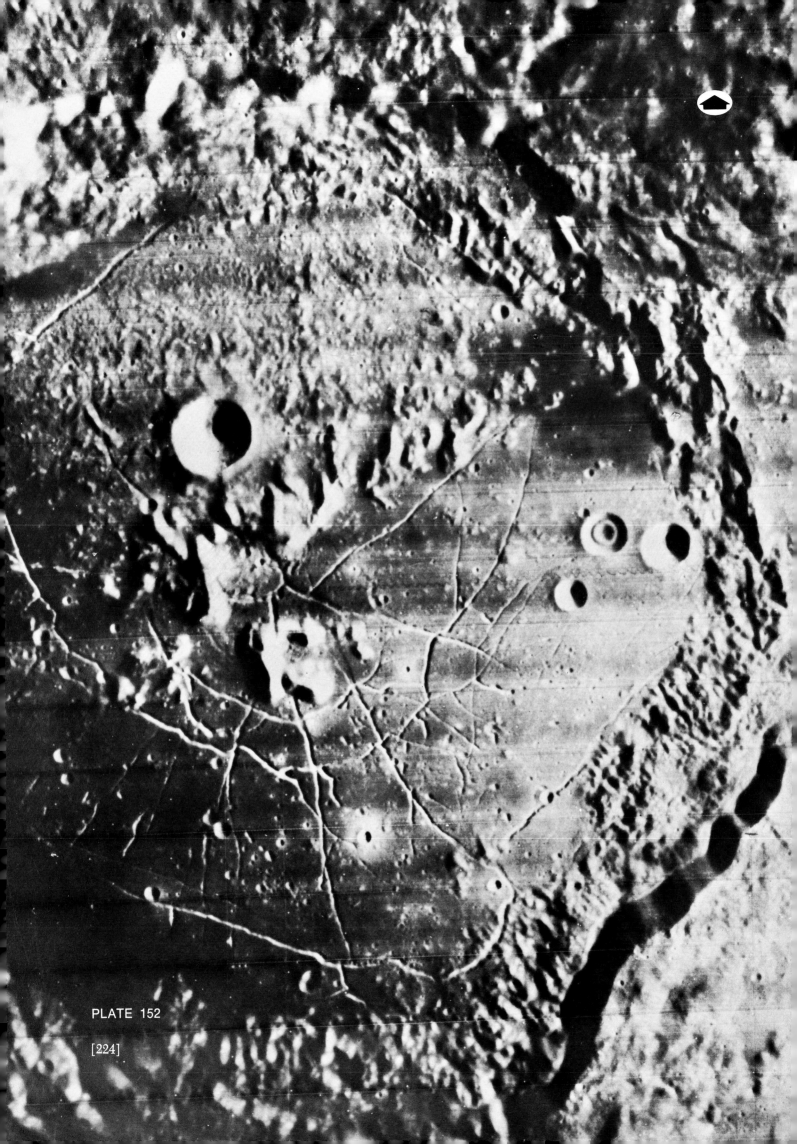

PLATE 152

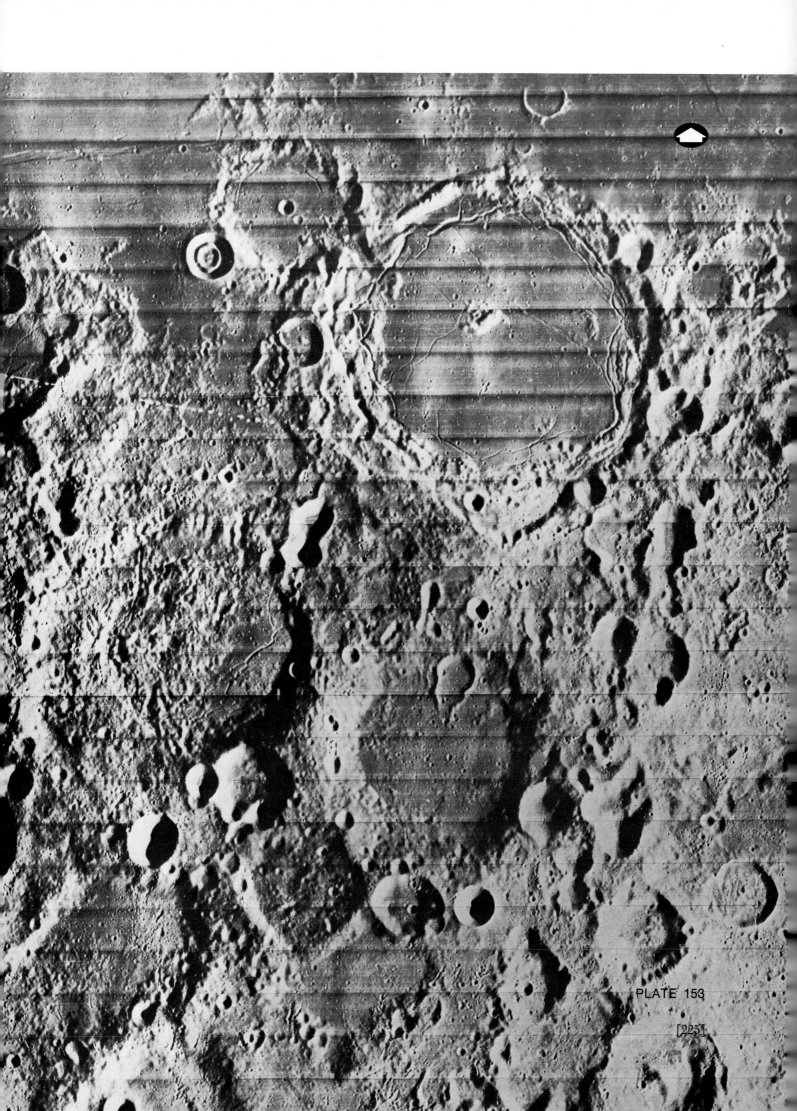

PLATE 153

[225]

PLATE 154

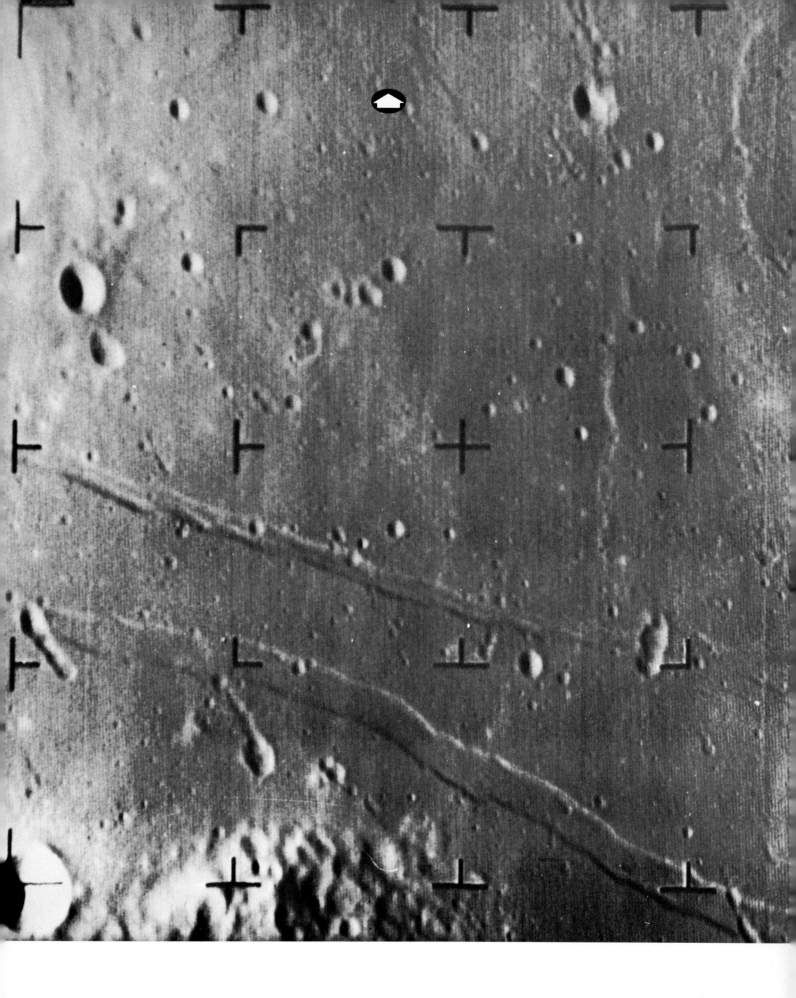

PLATE 155

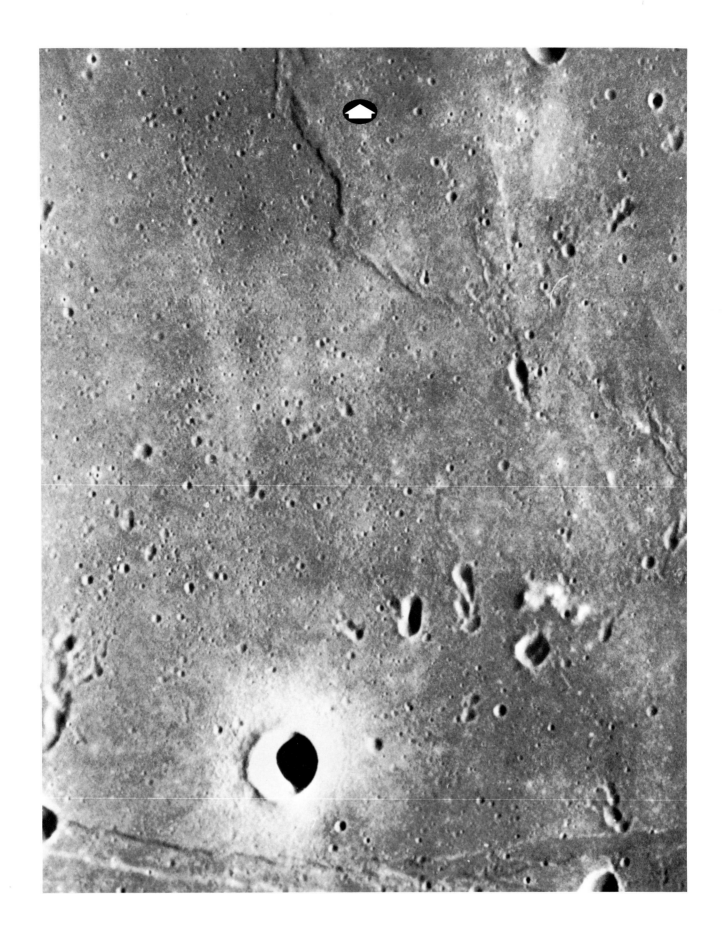

PLATE 156

PLATE 157

[229]

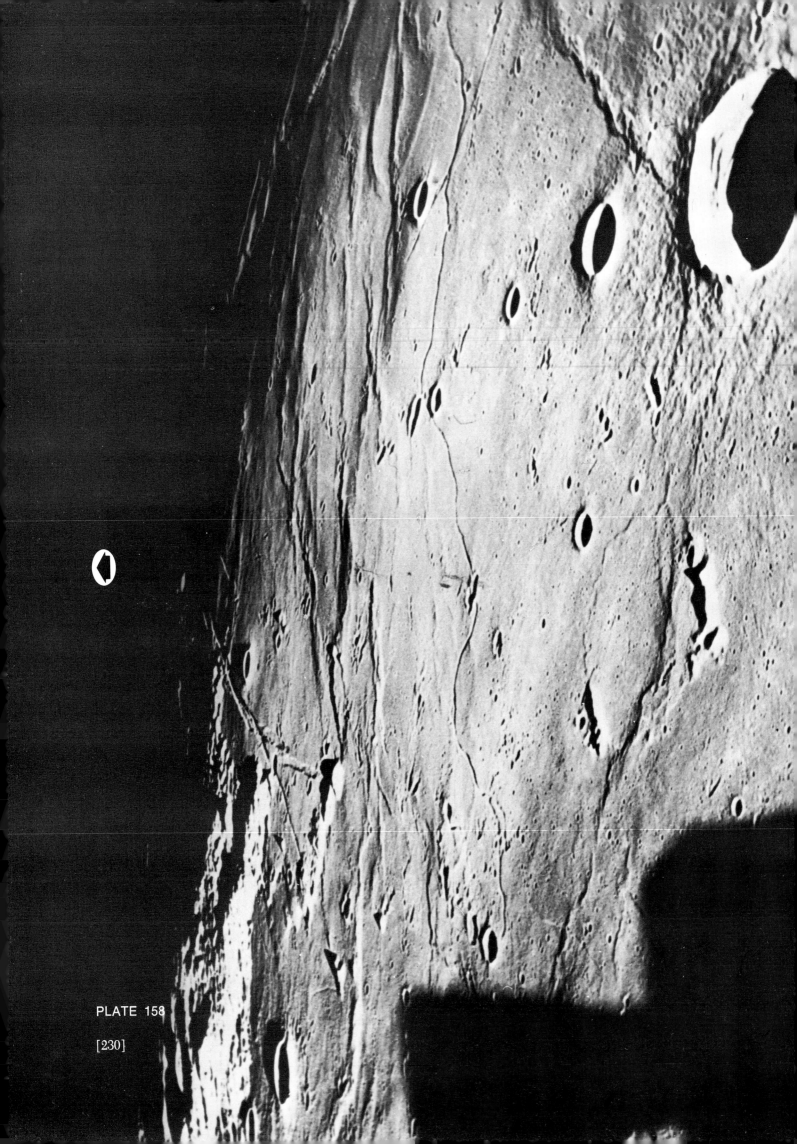

PLATE 158

PLATE 159

160. The "Cauchy hyperbolae" as seen from a different vantage point, as photographed by Apollo 8 on December 24, 1964 in the course of its circumlunar orbit. The two craters near the right margin of the frame are Taruntius D and E. (Reproduced by courtesy of NASA-Houston)

161. Sinus Medii and Mare Vaporum near the center of the apparent lunar disk visible from the Earth, with the crater Triesnecker (center) and its associated system of shallow rilles; the deeper Hyginus rille is to the northeast of it. Photograph taken on September 17, 1965 with the 43-inch reflector of the Observatoire du Pic-du-Midi. (Manchester Lunar Program)

162. Ariadeus rille between Mare Vaporum and Tranquillitatis, as photographed on May 26, 1964 with the 43-inch reflector of the Observatoire du Pic-du-Midi. (Manchester Lunar Program)

163. Sunrise over the Hyginus (left)and Ariadeus (center) rilles between Mare Vaporum and Tranquillitatis; the two craters (with central peaks) below the Ariadeus rille are Agrippa and Godin; above it is Julius Caesar. Photograph taken on August 4, 1965 with the 43-inch reflector of the Observatoire du Pic-du-Midi (Manchester Lunar Program)

164. The Hyginus rille in Mare Vaporum, as photographed by Bernard Lyot on March 21, 1945 with the 24-inch refractor of the Observatoire du Pic-du-Midi (below) and by the wide-angle lens of Lunar Orbiter 3 on 18 February 1967 from an altitude of 62 km above the lunar surface. (Reproduced by courtesy of NASA-Langley and of the Boeing Company)

165. An oblique view of the Hyginus rille from spacecraft, an enlargement of the same photograph as reproduced on Plate 163. (Reproduced by courtesy of NASA-Langley and of the Boeing Company)

166. A near-vertical view of the Hyginus rille, taken on August 14, 1967 by the wide-angle lens of Lunar Orbiter 5 from an altitude of 99 km. (Reproduced by courtesy of NASA-Langley and of the Boeing Company)

167. The Hyginus rille as photographed by Apollo 10 mission in May 1969 at the time of sunrise over Mare Vaporum. (Reproduced by courtesy of NASA-Houston)

168. Alpine valley (see also Plates 17 and 18), almost 150 km in length and up to 14 km in width, with a narrow rille running centrally through the most part of it. The depth of this rille ranges between 0.1–0.5 km, with a maximum near Mare Imbrium, where the cliffs flanking the valley attain an altitude in excess of 2,000 m. Photograph taken on May 20, 1967 by the high-resolution lens of Lunar Orbiter 4 from an altitude of 2,903 km. (Reproduced by courtesy of NASA-Langley and of the Boeing Company)

169. Eastern part of Mare Imbrium (Palus Putredinis), between the crater Archimedes (above) and the steepest part of the lunar Apennines (in the lower right corner) whose highest peak—Mount Bradley—attains an altitude close to 5,000 m. Note an extensive system of shallow wide rilles (akin to those of Sirsalis—Plate 181—or Hypatia—Plate 155 or 158), the widest part of which parallels the course of the Apennines. Photograph taken on May 18, 1967 with the high-resolution lens of Lunar Orbiter 4 from an altitude of 2,691 km. (Reproduced by courtesy of NASA-Langley and of the Boeing Company)

[232]

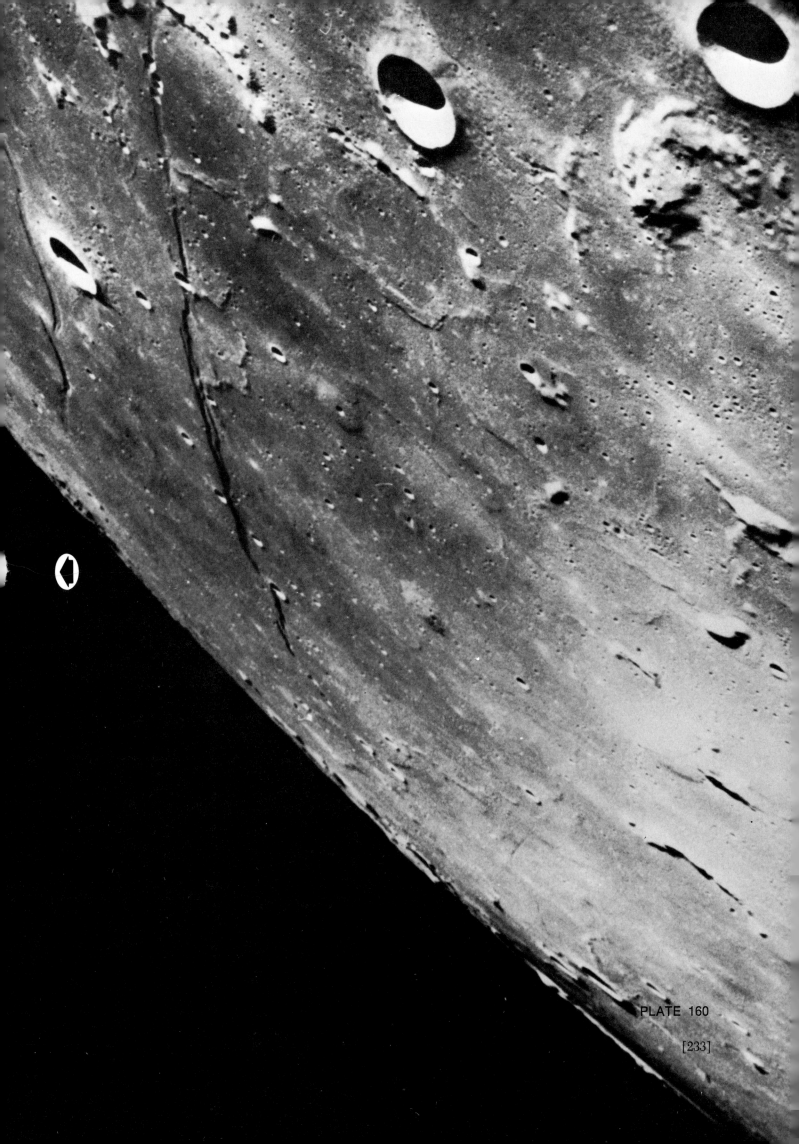

PLATE 160

[233]

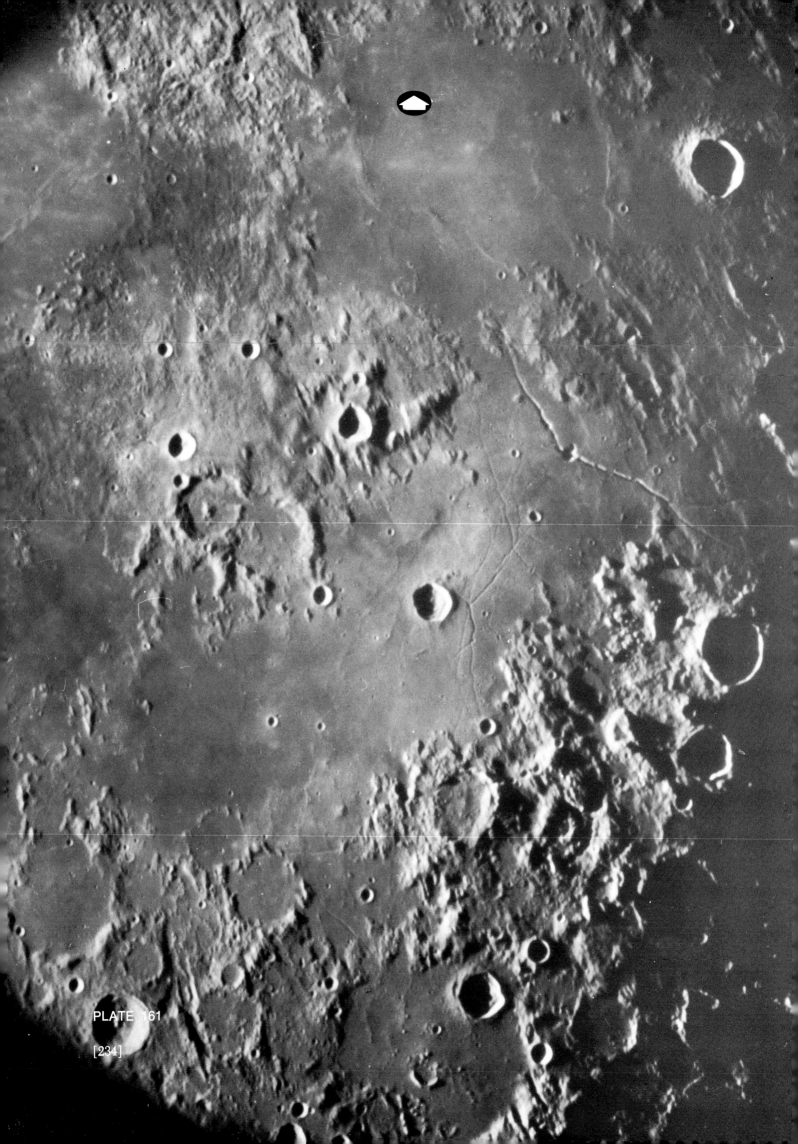

PLATE 161

PLATE 162

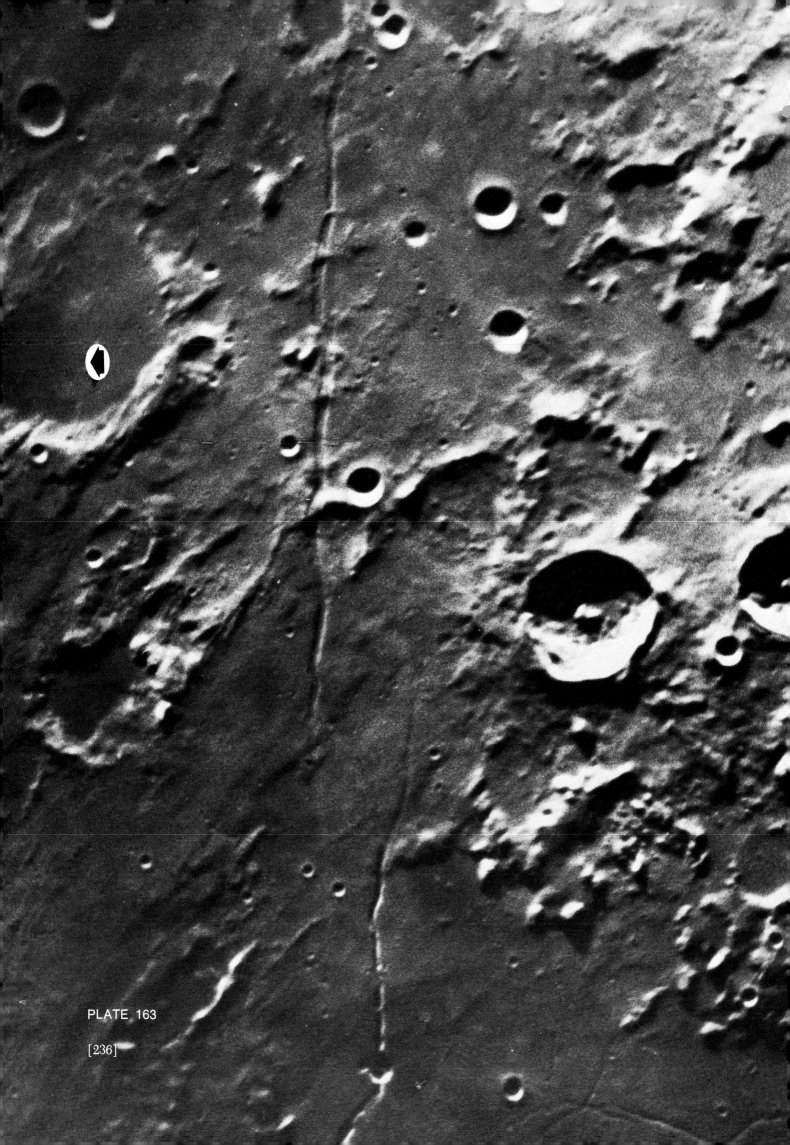

PLATE 163

[236]

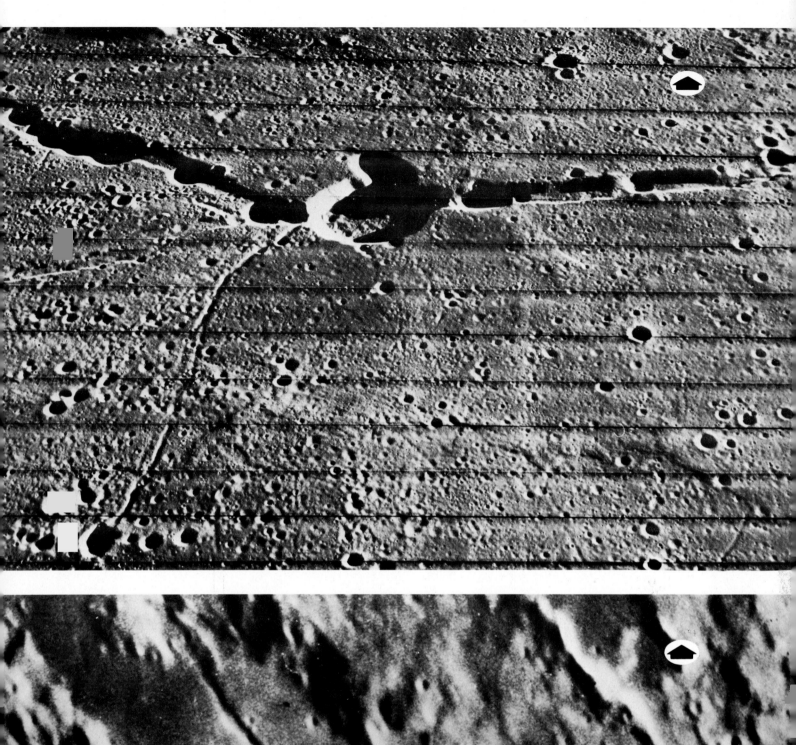

PLATE 164

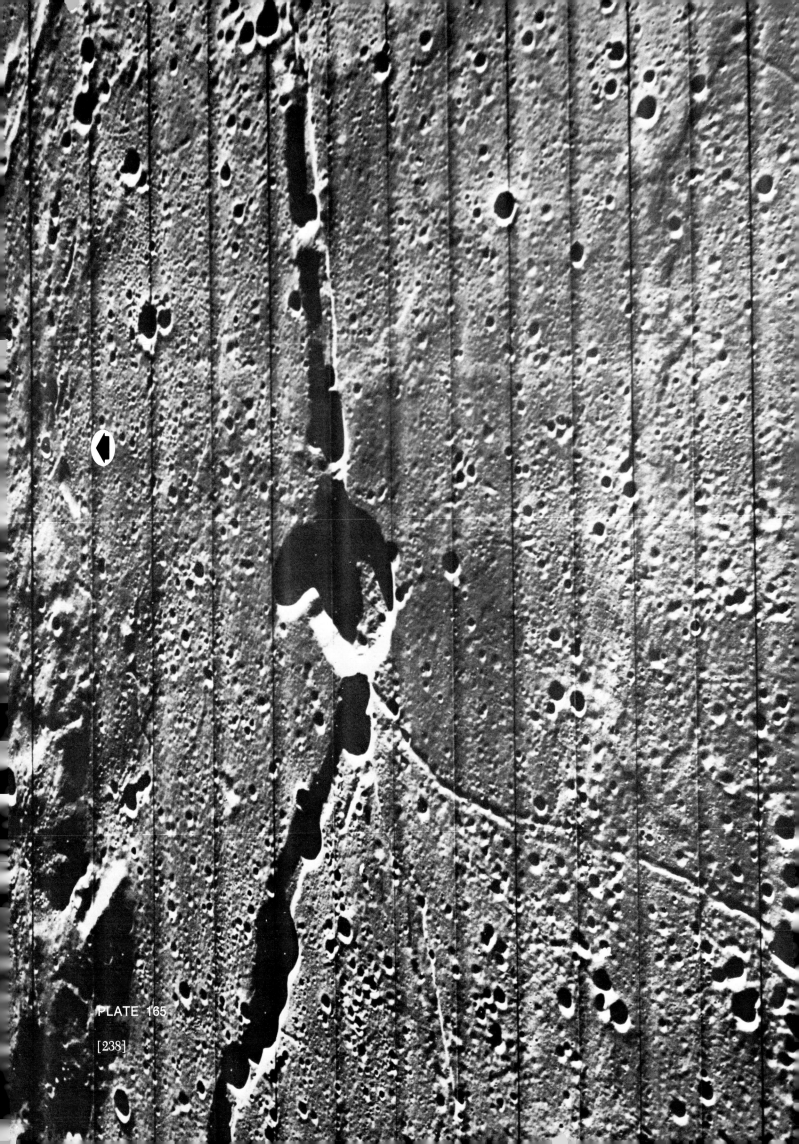

PLATE 165

[238]

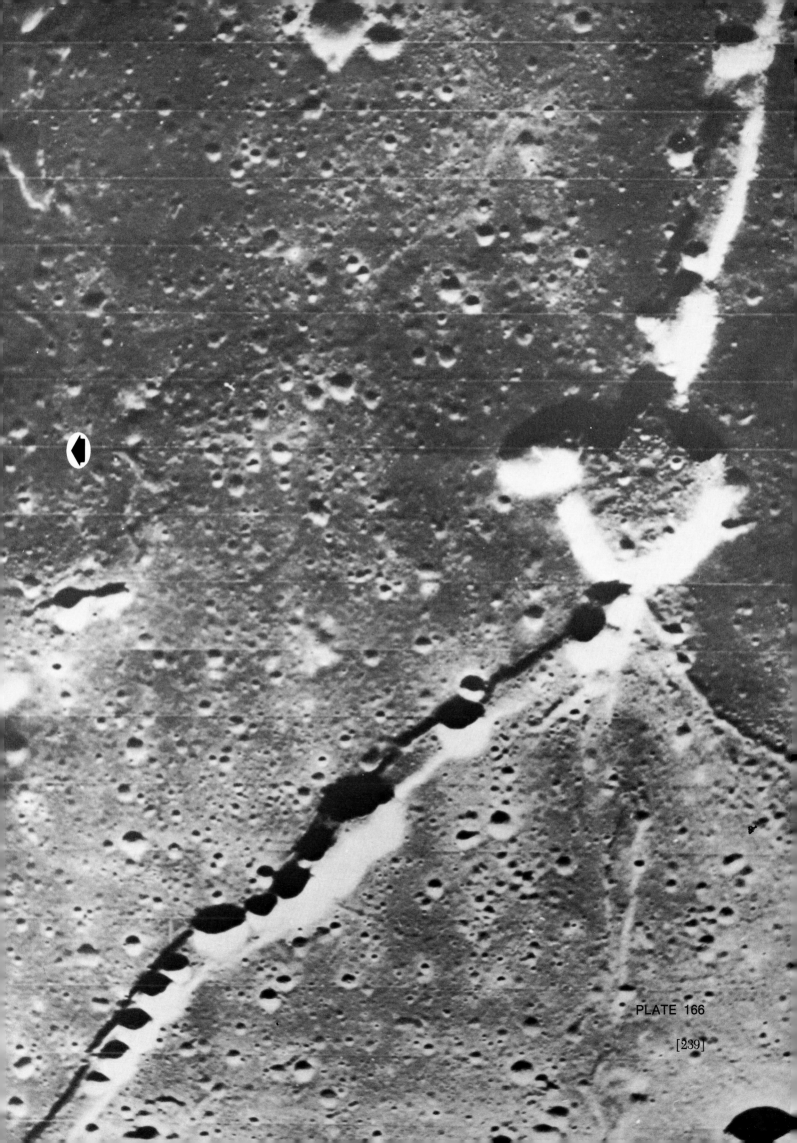

PLATE 166

[239]

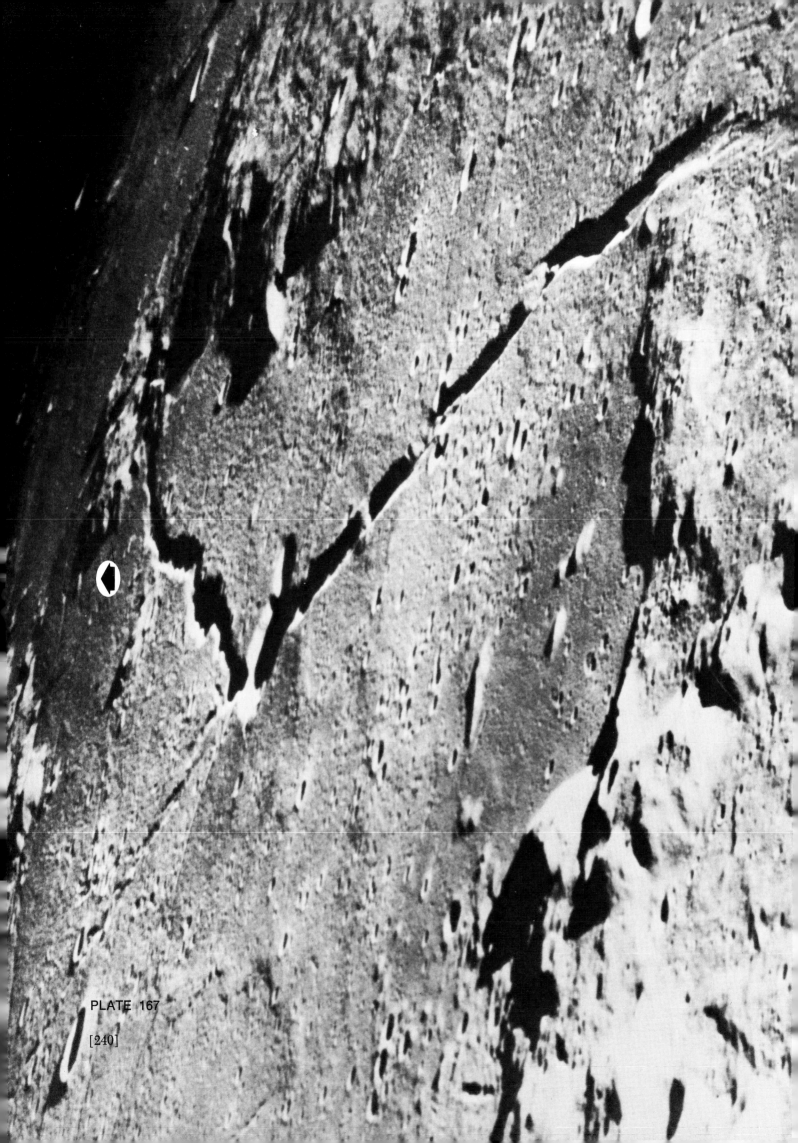

PLATE 167

[240]

PLATE 168

[241]

PLATE 169

170. A bird's-eye view of a part of the lunar Apennines, with the crater Conon (lower left) obviously produced by impact subsequent to the formation of the Apennine chain whose highest peak (Mount Bradley) rises immediately to the left of Conon. Above it are the plains of Palus Putredinis with a short but relatively deep Hadley rille (named after Mount Hadley, another one of the high Apennine peaks, overlooking its northern peak). Photograph taken on May 19, 1967 with the high-resolution lens of Lunar Orbiter 4 from an altitude of 2,697 km. Small white square denotes a section of this rille shown at high resolution on Plate 171. (Reproduced by courtesy of NASA-Langley and of the Boeing Company)

171. The close-up of the structure of a section of Hadley's rille seen on the preceding plate, as photographed on August 14, 1967 by the high-resolution lens of Lunar Orbiter 5 from an altitude of 129 km above the lunar surface (ground resolution, 2–3 m). (Reproduced by courtesy of NASA-Langley and of the Boeing Company)

172. The Bode rille, near the crater Bode in Sinus Aestuum south of the Apennines, as photographed on May 2, 1967 by the wide-angle lens of Lunar Orbiter 5 from an altitude of 105 km above the lunar surface. (Reproduced by courtesy of NASA-Langley and of the Boeing Company)

173. A high-resolution view of a part of Bode's rille, recorded by Lunar Orbiter 5 on May 2, 1967, only seconds after the photograph reproduced on the preceding plate was taken. Ground resolution: 2 to 3 m on the lunar surface. (Reproduced by courtesy of NASA-Langley and of the Boeing Company)

174. The crater Prinz and its associated system of rilles (see also Plates 24 and 25), as photographed on August 18, 1967 by the wide-angle lens of Orbiter 5 from an altitude of 136 km. For a high-resolution view of the section delimited by the white square see Plate 175. (Reproduced by courtesy of NASA-Langley and of the Boeing Company)

175. A high-resolution view of the section of one of Prinz's rilles (marked on the preceding plate by a white square), as recorded on August 18, 1967 by Lunar Orbiter 5 from an altitude of 136 km above the lunar surface. Ground resolution: 2 to 3 m. (Reproduced by courtesy of NASA-Langley and of the Boeing Company)

176–178 Photographs by Lunar Orbiter 5 of the Aristarchus-Herodotus region of the lunar surface with the well-known Schröter Canyon from the "cobra's head," at three increasing degrees of resolution. Photograph reproduced on Plate 176 was recorded on August 18, 1967 with the Lunar Orbiter's wide-angle lens from an altitude of 132 km; Plates 177 and 178 show telephoto views of diminishing sections of the rille (marked on Plates 176 and 177 with white squares) taken seconds later from an altitude of 134 km. (Reproduced by courtesy of NASA-Langley and of the Boeing Company)

179. Lunar landscape of Oceanus Procellarum west of Aristarchus and Herodotus (the area shown on Plate 124). The tail of Schröter Canyon is seen to its end on the right; the craters below it are Herodotus A, B, C; and the one dissected by the left margin is Schiaparelli. Photograph taken on May 22, 1967 by the high-resolution lens of Lunar Orbiter 4 from an altitude of 2,667 km. (Photograph reproduced by courtesy of NASA-Langley and of the Boeing Company)

PLATE 170

[244]

PLATE 171

[245]

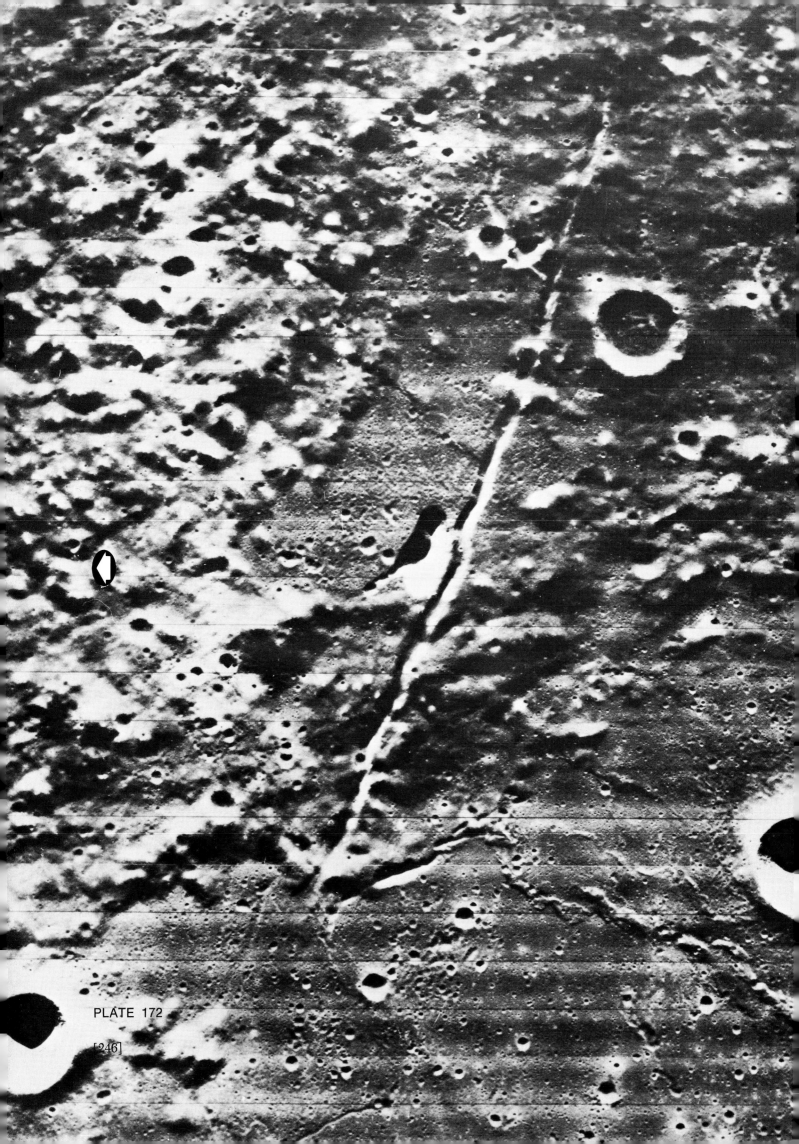

PLATE 172

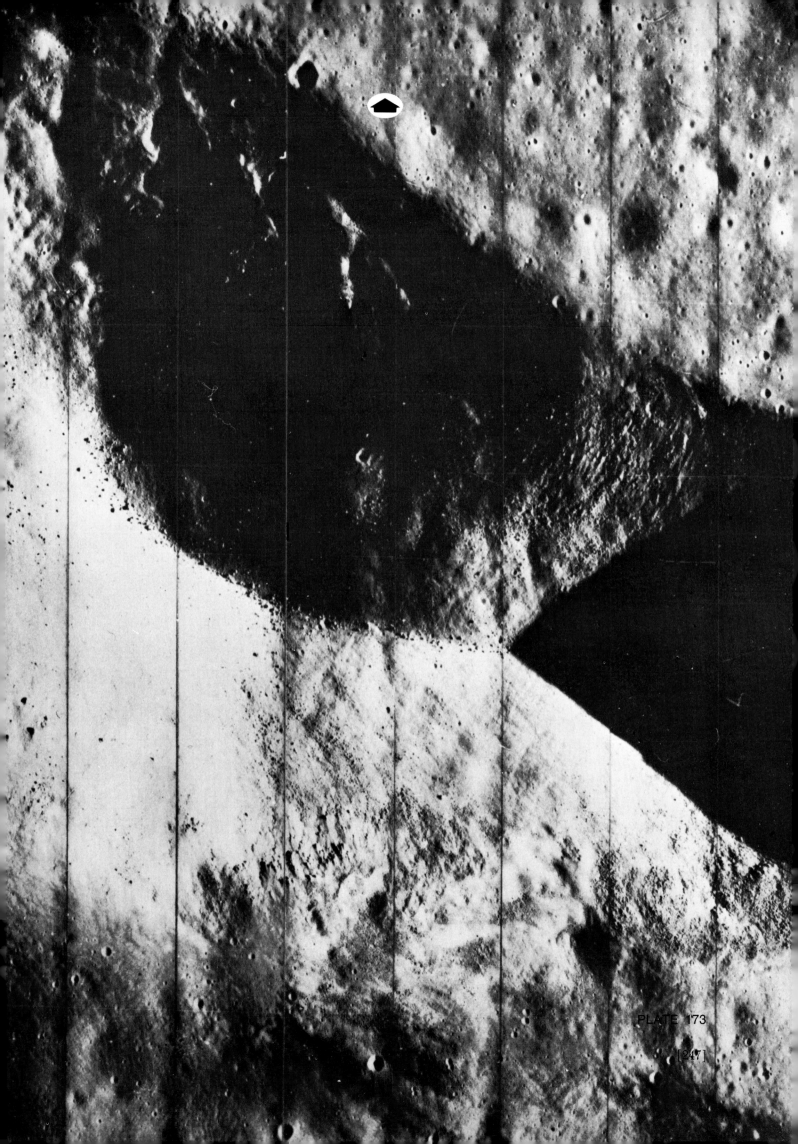

PLATE 173

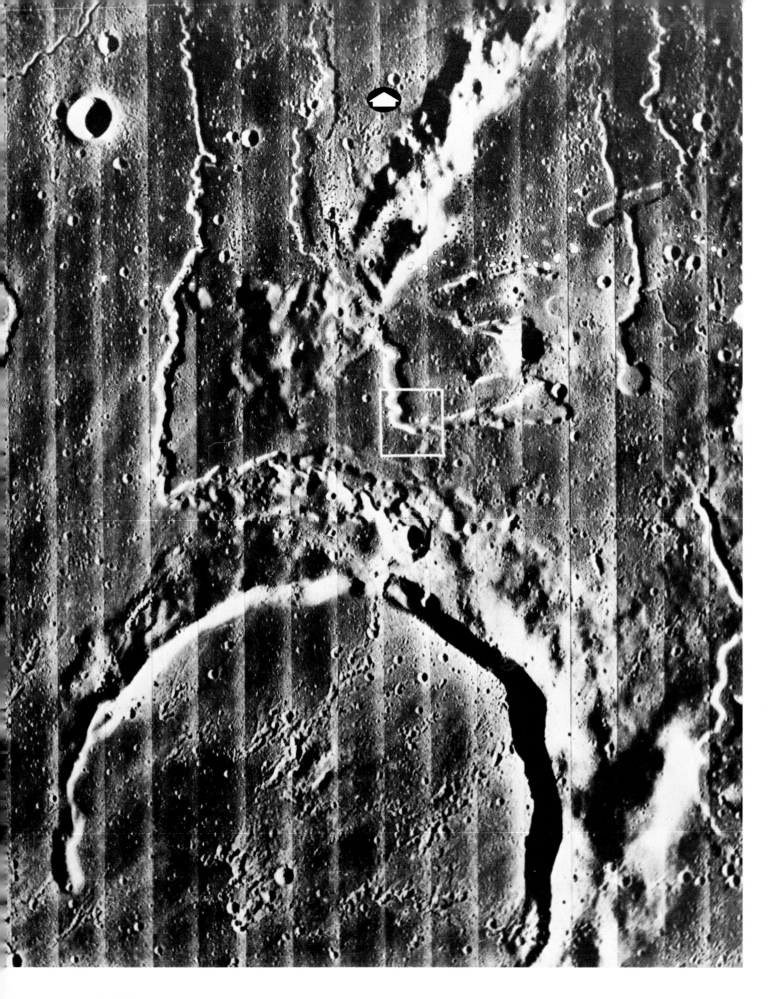

PLATE 174

[248]

PLATE 75

[249]

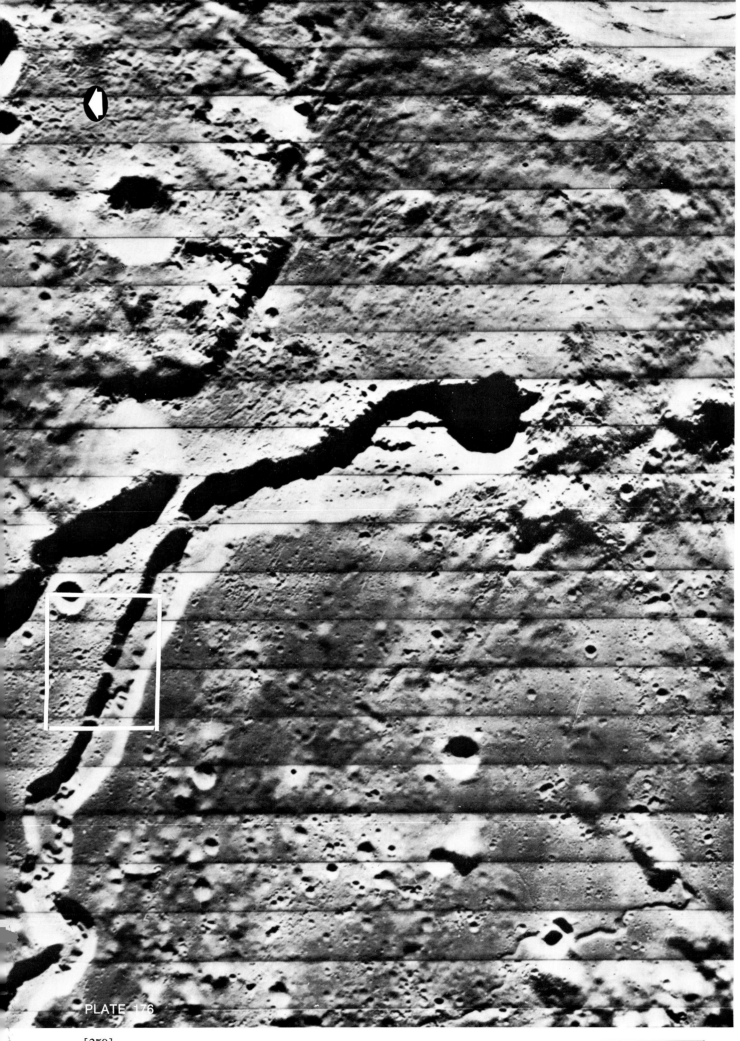

PLATE 176

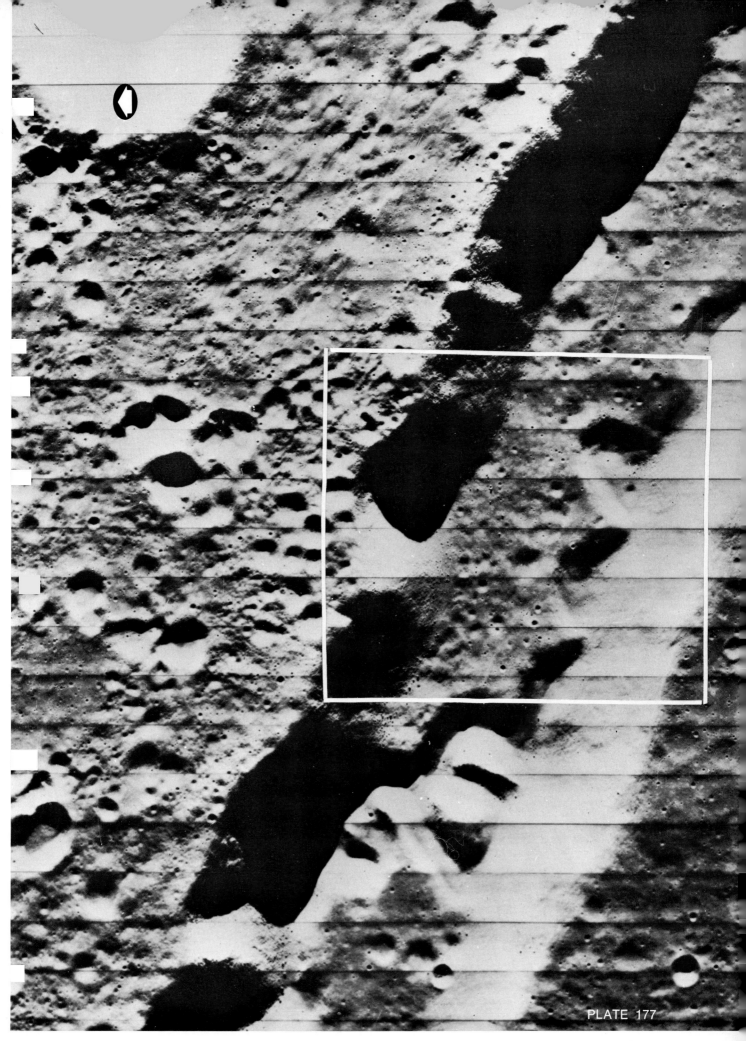

PLATE 177

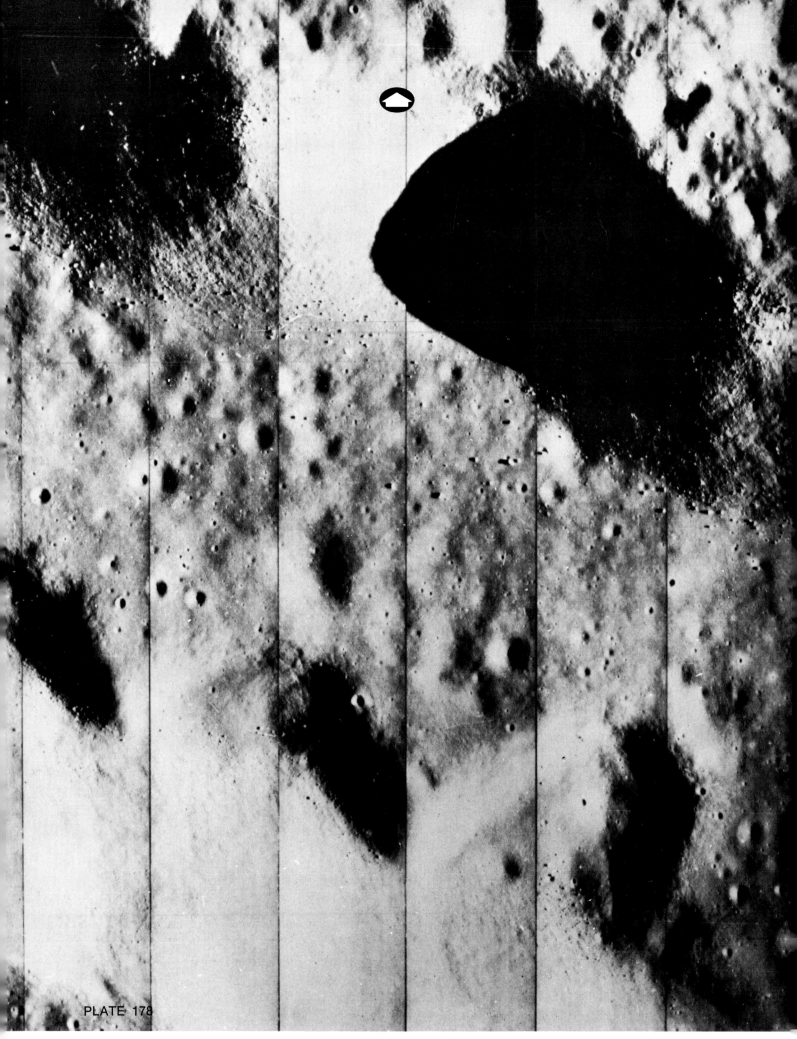

PLATE 178

PLATE 179

180. The Sharp rille on the western shores of Jura Mountains in Sinus Roris; the two craters near the right margin of the field are Mairan (middle) and Sharp (corner). The Sharp rille starts its meandering course south of Mairan as a shallow formation (akin to the rilles of Hypatia), but ends abruptly near the top as a canyon about 1.8 km deep. Photograph taken on May 22, 1967 by the high-resolution lens of Lunar Orbiter 4 from an altitude of 2,863 km. (Reproduced by courtesy of NASA-Langley and of the Boeing Company)

181. The Sirsalis rille near the west limb of the visible lunar hemisphere (east of Mare Orientale). A long shallow rille of the Hypatia type, traversing the walls of a number of craters, a fact evidencing that the rille was formed later than some of the craters (such as, for instance, De Vico A near the lower margin), but locally destroyed by other craters (such as Sirsalis J, just immediately below Sirsalis and Sirsalis A above). Moreover, several shallow rilles ("Darwin rilles") crisscross the Sirsalis rille in the lower left corner of the field—in a manner indicating that the Darwin rilles are older than Sirsalis. Photograph taken on May 23, 1967 by the high-resolution lens of Lunar Orbiter 4 from an altitude of 2,721 km. (Reproduced by courtesy of NASA-Langley and of the Boeing Company)

182. All previous photographs of the Moon reproduced in this atlas (with the exception of the one on Plate 13) were taken at the time of lunar sunrise or sunset, to bring into prominence the plastic relief of the respective parts of the surface by means of the shadows cast by any unevenness in the rays of the rising or setting sun. The next eight plates will illustrate the appearance of several parts of the lunar surface at full-moon conditions—at a time when no shadows are cast by the Sun standing high above the horizon and giving rise to enhanced constrast between the continental (that is, mountainous) and mare ground. The photograph shown on this plate shows high noon over Mare Nectaris and the southern bay of Mare Tranquillitatis, with the craters Theophilus and Mädler on its head (compare this photograph with one of the same region, reproduced on Plates 101 and 102). At high noon, lunar craters are discernible as bright rings by greater reflectivity of the rims of their ramparts, rather than by any shadows. Photograph taken on October 9, 1965 with the 61-inch reflector of the U.S. Naval Observatory at Flagstaff, Arizona. (Reproduced by courtesy of A.C.I.C.)

183. Lunar noon over the eastern part of Mare Serenitatis (left), with the crater Posidonius (top) and Proclus spreading its fanlike bright rays in the lower right corner of the field. Photograph taken on October 9, 1965 with the 61-inch reflector of the U.S. Naval Observatory at Flagstaff, Arizona. (Reproduced by courtesy of A.C.I.C.)

184. The eastern part of Mare Frigoris and a view toward the north pole of the Moon as it appears near full moon. The outlines of the crater Aristoteles can be seen near the lower margin of the field; the ray craters of Anaxagoras can be seen near the upper left corner of the frame; Thales is near the upper right. Photograph taken on October 9, 1965 with the 61-inch reflector of the U.S. Naval Observatory at Flagstaff, Arizona. (Reproduced by courtesy of A.C.I.C.)

185. Mare Serenitatis (right) and the lunar Alps and Caucasus under noon illumination. Of the principal craters visible in the field, Aristillus can be seen in the lower left part of the frame; the slender ring above it is Cassini. To the right we see Aristoteles and Eudoxus; while the Alpine Valley can be seen in the upper left corner. Photograph taken October 9, 1965 with the 61-inch reflector of the U.S. Naval Observatory at Flagstaff, Arizona. (Reproduced by courtesy of A.C.I.C.)

PLATE 180

PLATE 181

[256]

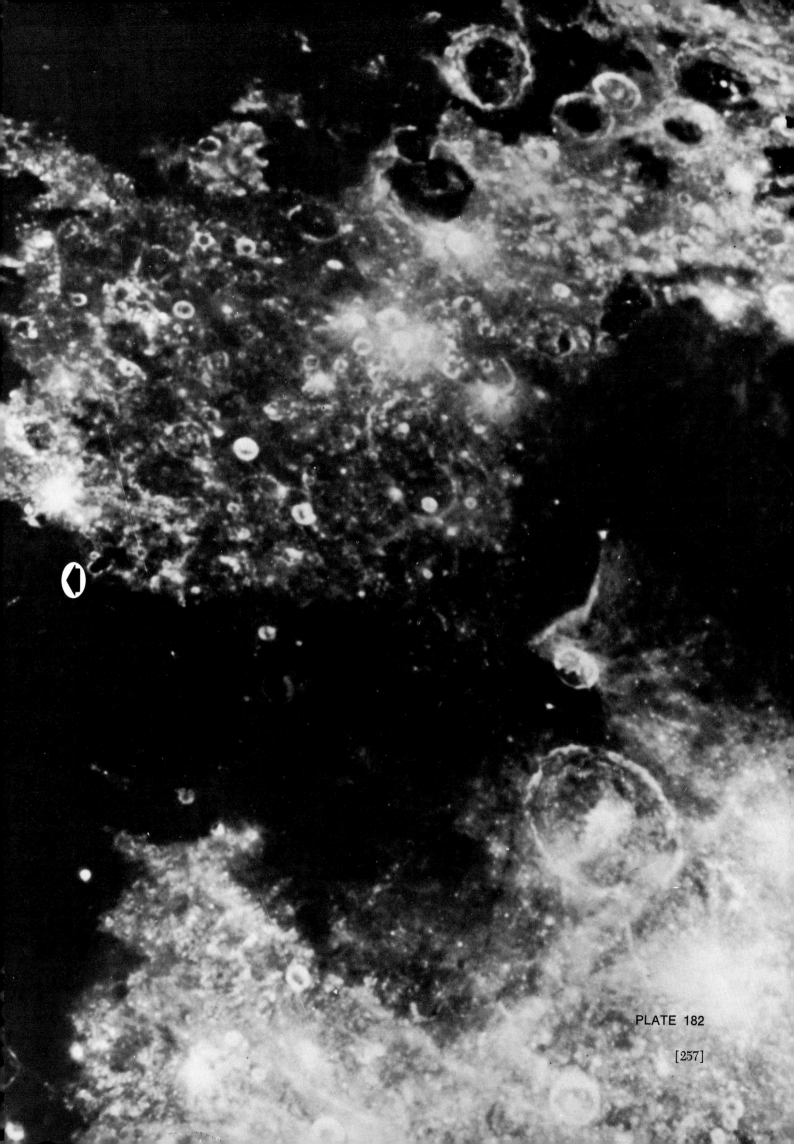

PLATE 182

[257]

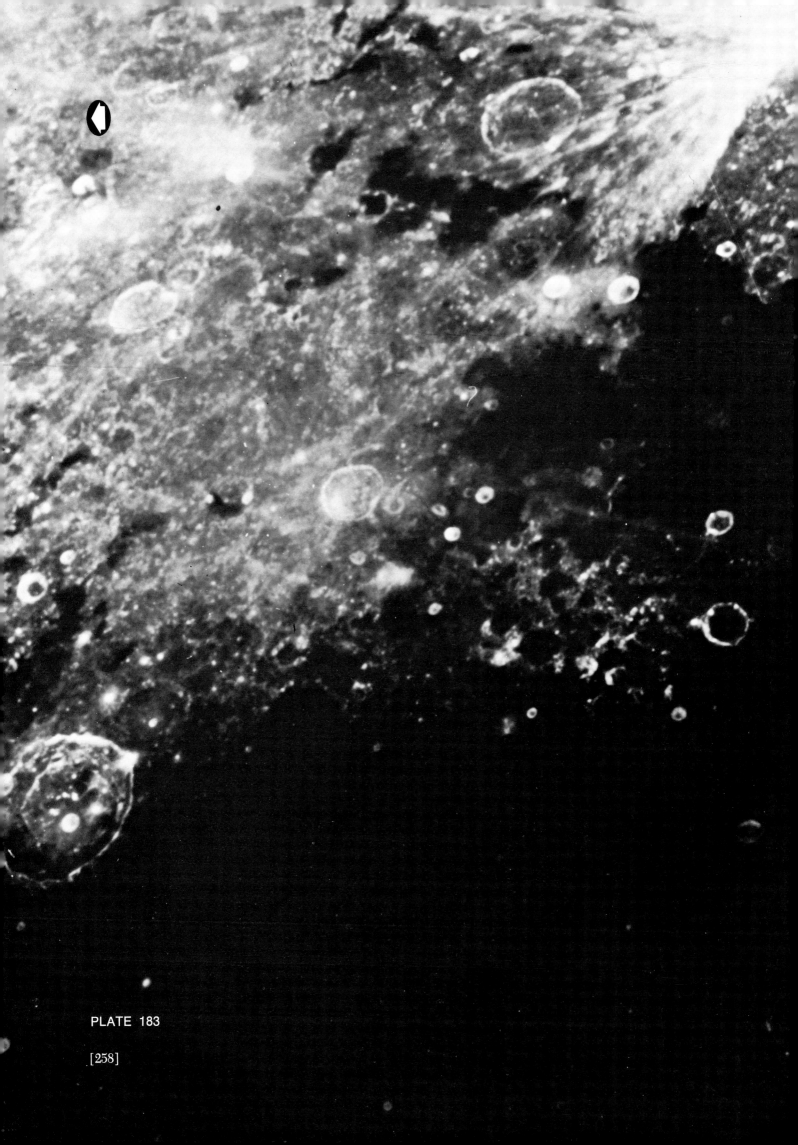

PLATE 183

[258]

PLATE 184

[259]

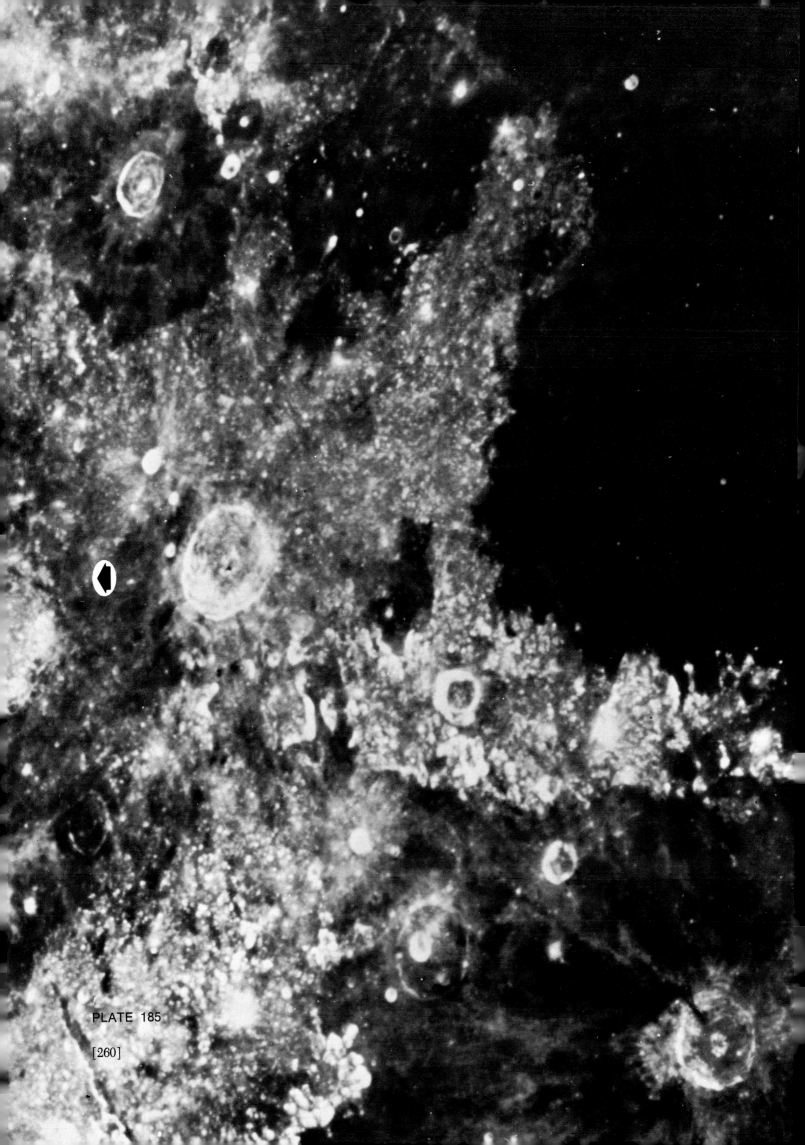

PLATE 185

[260]

186. Western part of Mare Frigoris (near the Moon's north pole) with the craters Plato (dark-floor circular formation, below) and Anaxagoras (bright ray, upper right). Photograph taken on October 9, 1965 with the 61-inch reflector of the U.S. Naval Observatory at Flagstaff, Arizona. (Reproduced by courtesy of A.C.I.C.)

187. Northern part of Mare Imbrium at high noon, with the craters Aristillus and Cassini (lower right) and Plato (top); the Alpine Valley can be seen in between. Photograph taken on October 9, 1965 with the 61-inch reflector of the U.S. Naval Observatory at Flagstaff, Arizona. (Reproduced by courtesy of A.C.I.C.)

188. Southern part of Mare Imbrium with the lunar Apennines and the crater Archimedes on the right, and the Copernican bright rays protruding from the lower left corner. Photograph taken on October 9, 1965 with the 61-inch reflector of the U.S. Naval Observatory at Flagstaff, Arizona. (Reproduced by courtesy of A.C.I.C.)

189. Sinus Aestuum (center) and the crater Copernicus with its extensive system of divergent bright rays to the left. Note the presence of several small "dark-haloed" craters on the Copernican white apron. Photograph taken on October 9, 1965 with the 61-inch reflector of the U.S. Naval Observatory at Flagstaff, Arizona. (Reproduced by courtesy of A.C.I.C.)

190. Close-up of the landing places of Ranger 7 (upper right) on July 31, 1964); of Ranger 8 (upper left) on February 20, 1965; and of Ranger 9 (below) on March 24, 1965—all televised from altitudes 4–6 km above the lunar surface, only seconds before impact. The size of the three fields is approximately 2.5 km across on the Moon; and the smallest details resolved on them are of the order of 1 m for Rangers 7 and 8, and substantially smaller for Ranger 9. On this resolution, the surface structure of all three regions appears to be essentially the same, suggesting that the processes which shaped it are global, rather than local, in nature. (Reproduced by courtesy of NASA-JPL)

191. The landing place of the Surveyor 1 in the Oceanus Procellarum near the crater Flamsteed, as photographed by Lunar Orbiter 1 on August 29, 1966 (left); the spacecraft itself was located by its shadow on an Orbiter 3 photograph taken on February 22, 1967 from an altitude of about 52 km (right). The photograph below shows ramparts of the 1-km crater, recorded by Orbiter 1 from above, as televised from the lunar ground by the Surveyor itself. (Reproduced by courtesy of NASA, JPL and the Boeing Company)

192. One of the footpads of the Surveyor 1, which sank on touchdown on June 4, 1966 several centimeters into the lunar ground (indicative of the surface bearing strength of a few pounds per square inch). (Reproduced by courtesy of NASA-JPL)

193. The immediate surroundings of one of the landing footpads of the Surveyor spacecraft is shown on this composite picture of several images televised by the Surveyor's high-resolution optical system on June 13, 1966. The evidence discloses vividly the nature of the rough ground (consisting of clods of soil) seen at a range of less than 1½ m. At the time of the exposure, the Sun stood about 10° above the lunar horizon; and the smallest details resolvable on the photograph are less than 1 mm in size. (Reproduced by courtesy of NASA-JPL)

194. Further details of the lunar ground shown on the preceding plate.

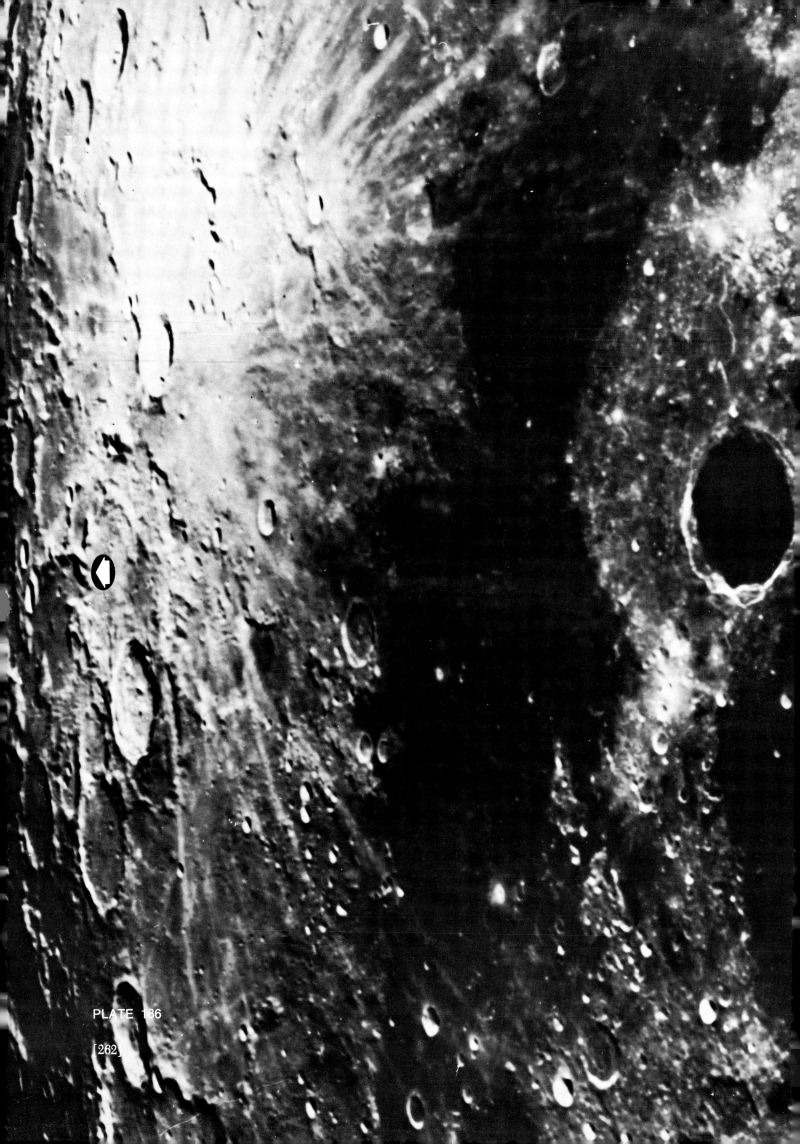

PLATE 186

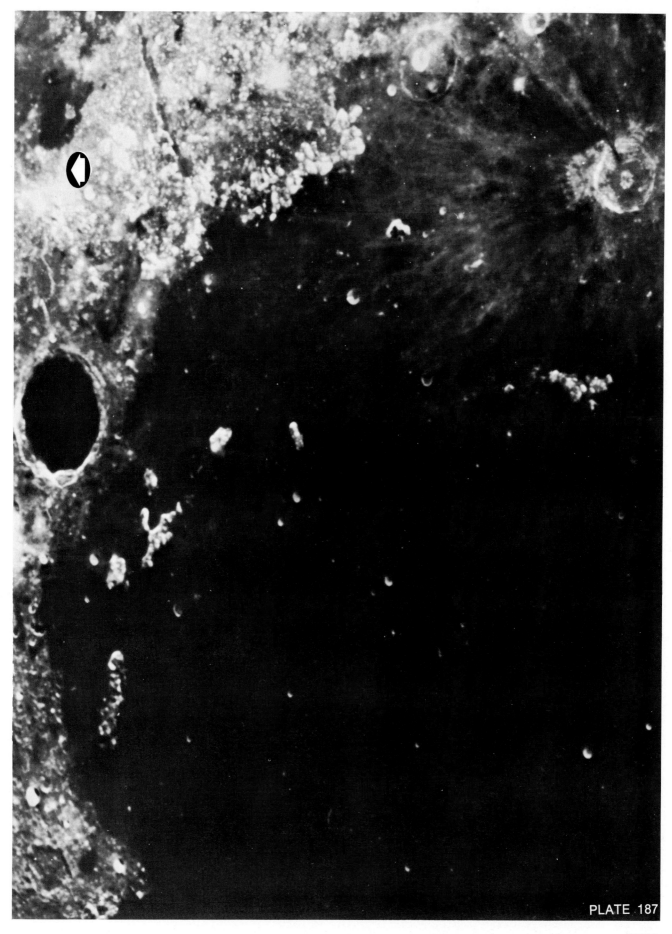

PLATE 187

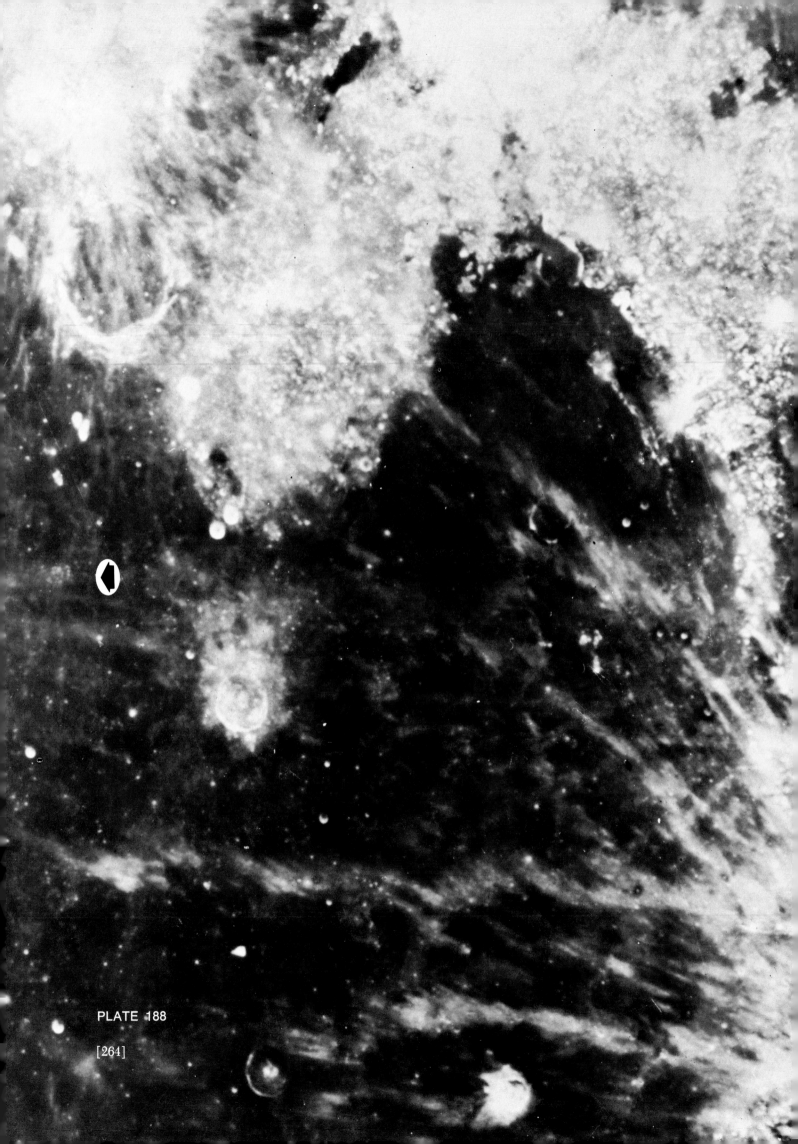

PLATE 188

[264]

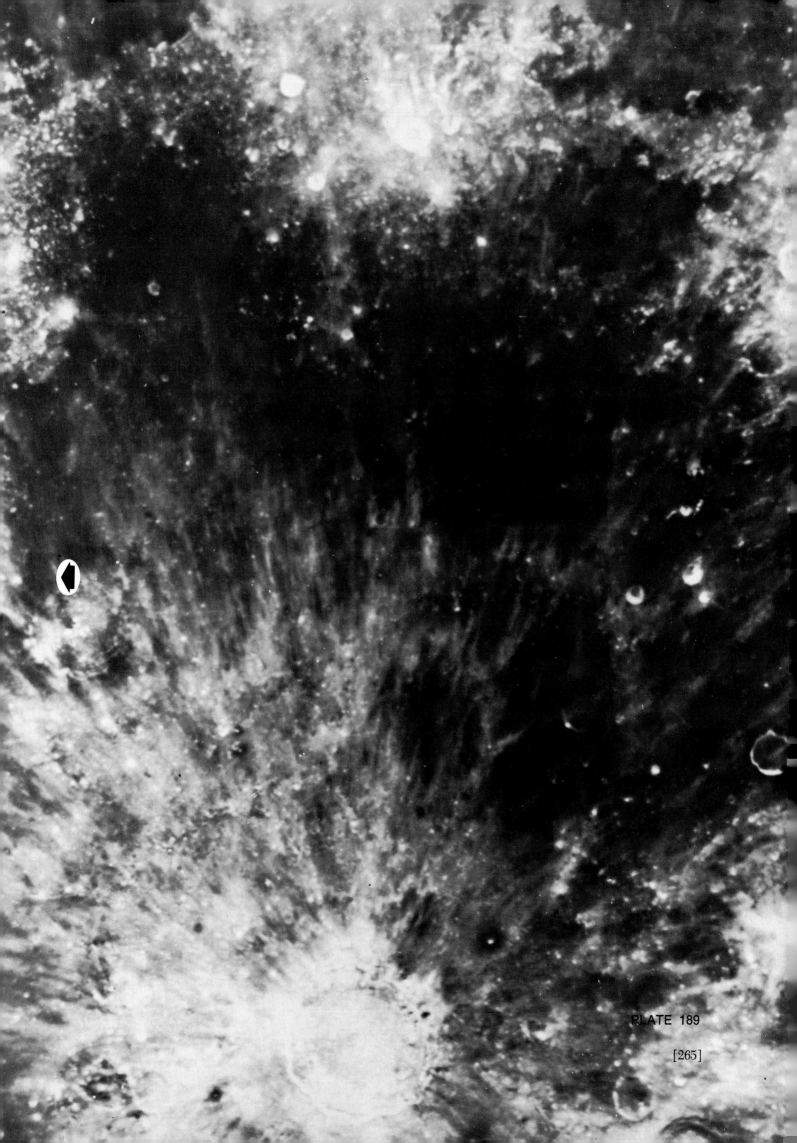

PLATE 189

[265]

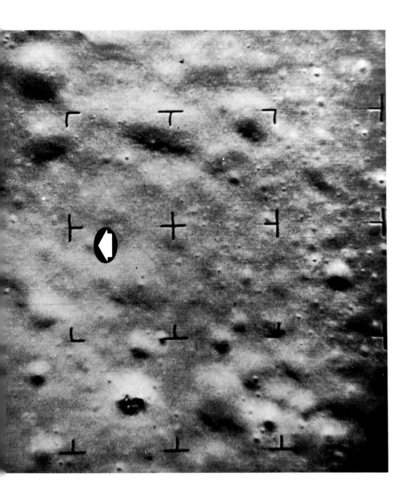

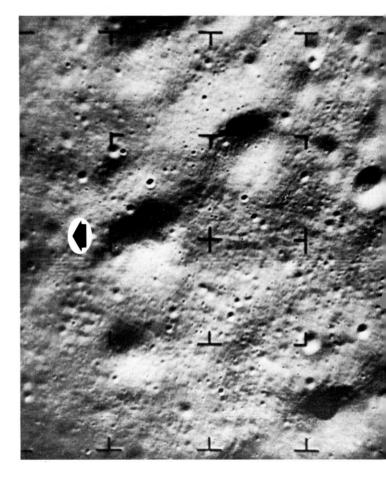

PLATE 150

[266]

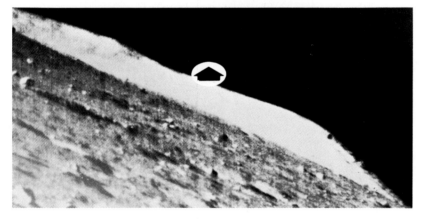

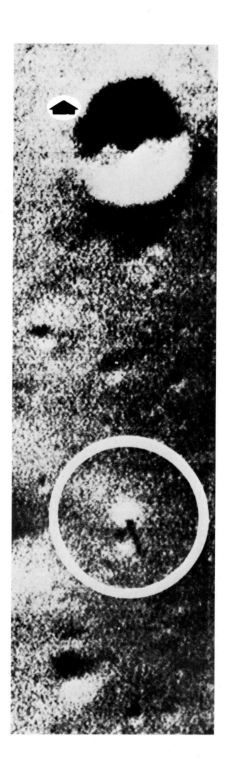

PLATE 191

PLATE 192

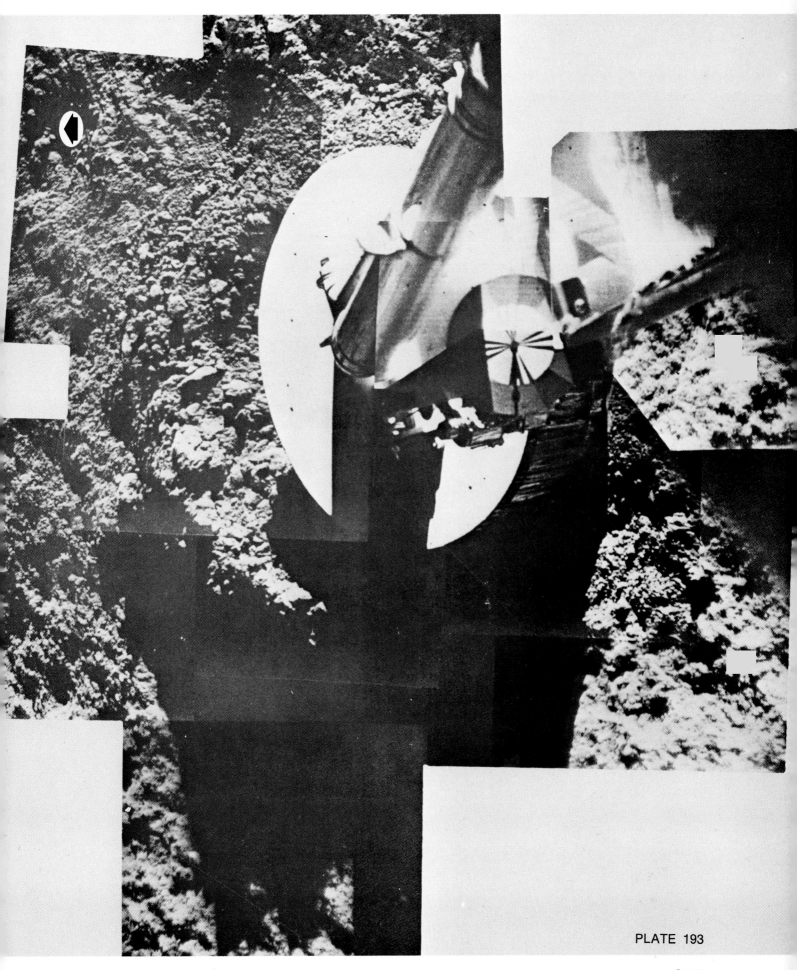

PLATE 193

PLATE 194

[270]

195. Late lunar afternoon over the surface in Oceanus Procellarum televised to us by Surveyor 1 on June 12, 1966. Note the flat horizontal panorama of lunar landscape, and a small crater—probably of secondary impact origin—some 3 m in size, at a distance of about 26 m from the spacecraft. (Reproduced by courtesy of NASA-JPL)

196. Sunset over the lunar landscape in Oceanus Procellarum (near the crater Flamsteed) as televised to the Earth by Surveyor 1 on June 14, 1966; the dark regions are already covered by shadows. (Reproduced by courtesy of NASA-JPL)

197. A close-up view of a portion of the lunar surface near the time of sunset, as televised by Surveyor 1 on June 13, 1966. Note the long shadow cast by a small boulder in the immediate neighborhood of the spacecraft. Smallest details discernible on the photograph are only millimeters in size. (Reproduced by courtesy of NASA-JPL)

198. A part of the horizon of the lunar surface, televised to the Earth by Luna 9 on February 4, 1966 from its landing place in the western parts of the Oceanus Procellarum near the crater Cavalerius. The stones in the foreground are 15–20 cm in size, and lie about 2–3 m from the camera. The smallest details seen in the proximity of the spacecraft are only a few millimeters in size. (Reproduced by courtesy of the USSR Academy of Sciences)

199. Composite horizontal panorama of a more rugged continental part of the lunar surface northeast of the landing place of Surveyor 7 (about 29 km north of the crater Tycho). The horizon (made up of a series of hills and ridges) is about 13–20 km away. (Image televised on January 10, 1968, and reproduced by courtesy of NASA-JPL)

200. A close-up of the mountainous and stony ground in the proximity of the landing place of Surveyor 7, as televised by the spacecraft's high-resolution optics. (Reproduced by courtesy of NASA-JPL)

201. Horizontal panorama as seen from Tranquility Base on July 20, 1969—astronaut Neil Armstrong took this picture on the spot. (Reproduced by courtesy of NASA-Houston)

202. An area of approximately 250×170 m of the lunar surface near the Ariadaeus rille (centered on $\lambda = 15°\!.3$ E, $\beta = 4°\!.5$ N) as recorded by the telephoto lens of Lunar Orbiter 2 on November 7, 1966 from an altitude of 49 km, showing examples of some of the largest individual boulders photographed anywhere on the Moon. The white cross mark on the plate is approximately 8×8 m in size. (Reproduced by courtesy of NASA-Langley and of the Boeing Company)

203. Boulders covering the slopes of shallow craters: a Lunar Orbiter 3 photograph taken with the spacecraft's high-resolution optics on February 15, 1967 from an altitude of 51 km above the lunar surface. The field of the plate covers an area of approximately 1.0×1.2 km on the lunar surface, in a position described by the selenographic coordinates $\lambda = 35°\!.11$ E and $\beta = 2°\!.35$ N. (Reproduced by courtesy of NASA-Langley and of the Boeing Company)

PLATE 195

[272]

PLATE 196

[273]

PLATE 197

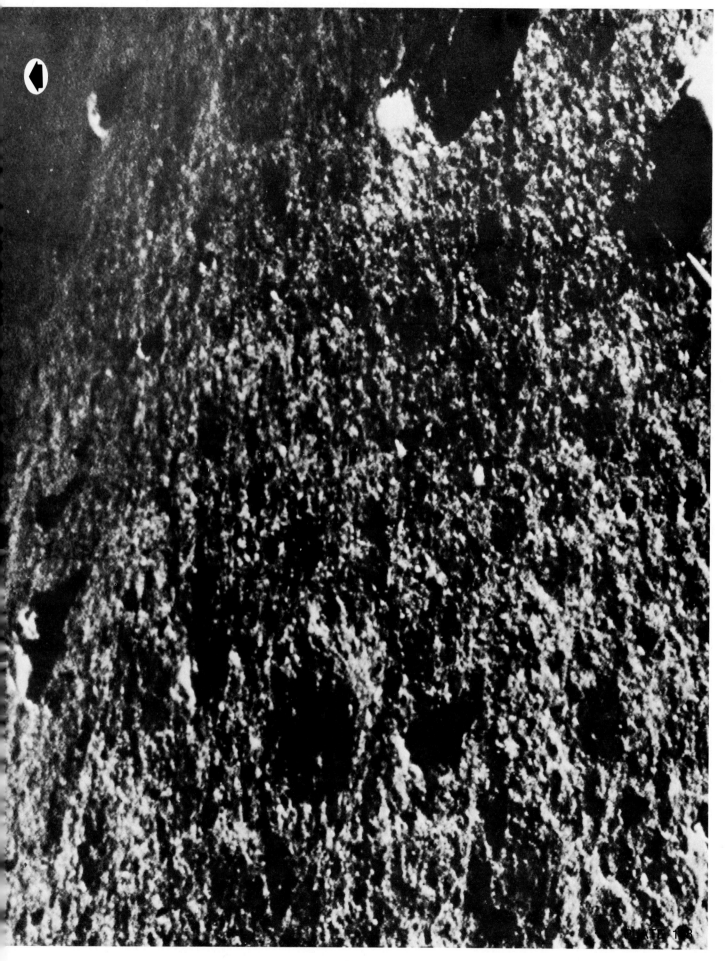

PLATE 1.8

PLATE 199

PLATE 200

[277]

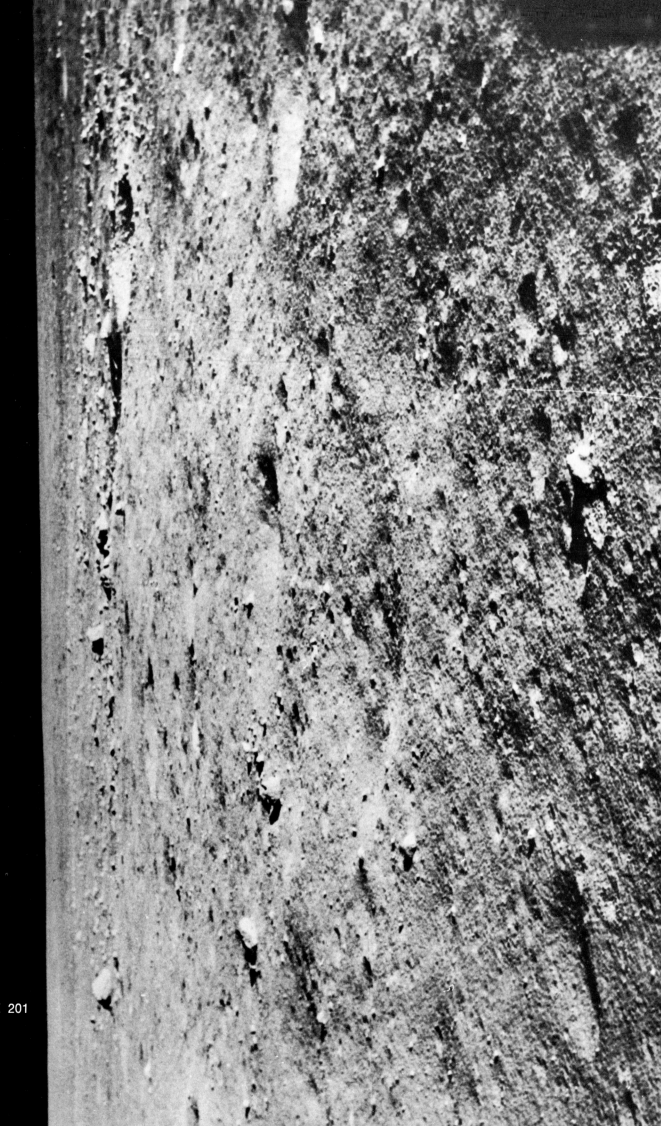

PLATE 201

[278]

PLATE 202

[279]

PLATE 203

204. A small well-defined crater of recent origin, situated in Oceanus Procellarum at an approximate selenographic position specified by $\lambda = 46°1$ W, $\beta = 2°0$ N. The field shown on this plate is approximately 1.1×0.8 km in size, and the diameter of the crater is 140 m. Its double-walled appearance is probably due to a continuous landslide around the circumference of the crater, occurring shortly after the crater was formed. Boulders up to a few meters in size were ejected from it at that time to form a symmetrical pattern of rays. Photograph taken on February 21, 1967 with the telephoto lens of Lunar Orbiter 3 from an altitude of 52 km. (Reproduced by courtesy of NASA-Langley and of the Boeing Company)

205. A Lunar Orbiter 5 photograph of lunar ground in the neighborhood of the crater Vitello showing several tracks of rolling boulders, taken on August 17, 1967 from an altitude of 167 km. The approximate lunar coordinates of the field's center are $\lambda = 37°61$ W, $\beta = 30°80$ S; framelet width, 710 m, and ground resolution close to 2 m. (Reproduced by courtesy of NASA-Langley and of the Boeing Company)

206. An enlarged view of the boulder track seen at the extreme left of Plate 205, showing the sinuous trail in the ground left behind the rolling boulder resting at the end of it. The depth of this trail (which can be determined from shadow measurements) can disclose the bearing strength of the relatively soft surface over which the boulder has rolled down to its resting place. (Reproduced by courtesy of NASA-Langley and of the Boeing Company)

207. A Lunar Orbiter 5 photograph of the foot of one of the hills forming the central mountain of the crater Copernicus, taken with the high-resolution optics of the spacecraft on August 16, 1967 from an altitude of 103 km. The approximate lunar coordinates of the field's center are $\lambda = 20°31$ W, $\beta = 9°71$ N, and the ground resolution is less than 2 m. (Reproduced by courtesy of NASA-Langley and of the Boeing Company)

208. A group of small domes, with boulder fields perched on their tops, on the floor of Copernicus northwest of its central mountain. The resolving power of this telephoto frame taken by Lunar Orbiter 5 on August 16, 1967 is between 1–2 m on the lunar surface. (Reproduced by courtesy of NASA-Langley and of the Boeing Company)

209. A group of small lunar domes on the floor of the crater Copernicus, northwest of its central hills, showing groups of boulders perched precariously on their tops—a feature which has earned such domes the playful name of "lunar hedgehogs." The size of the individual boulders seen in great numbers on their exposed positions ranges between 2–15 m. Photograph taken on August 16, 1967 by the high-resolution lens of Lunar Orbiter 5 from an altitude of 102 km. White rectangle marks the area shown in enlargement on Plate 210. (Reproduced by courtesy of NASA-Langley and of the Boeing Company)

210. An enlargement of a part of the photograph reproduced on the preceding plate (marked on the latter by a white square), showing details of one particular lunar hedgehog on the floor of Copernicus. The resolving power of this telephoto frame taken by Orbiter 5 on August 16, 1967 is close to 1 m on the lunar surface. (Reproduced by courtesy of NASA-Langley and of the Boeing Company)

PLATE 204

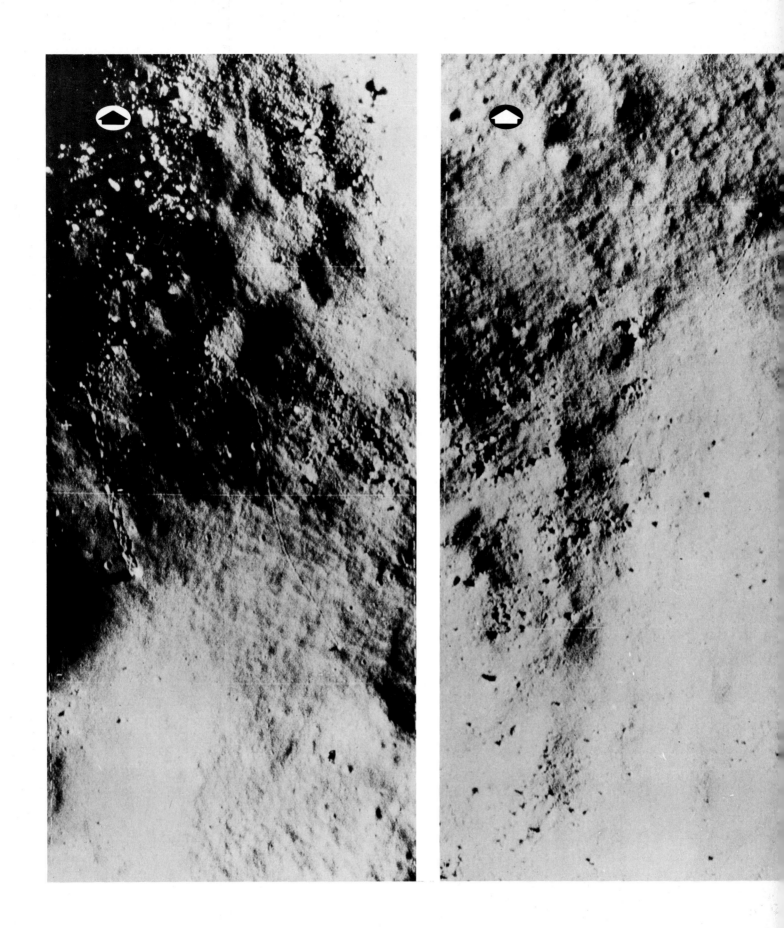

PLATE 205

PLATE 206

[284]

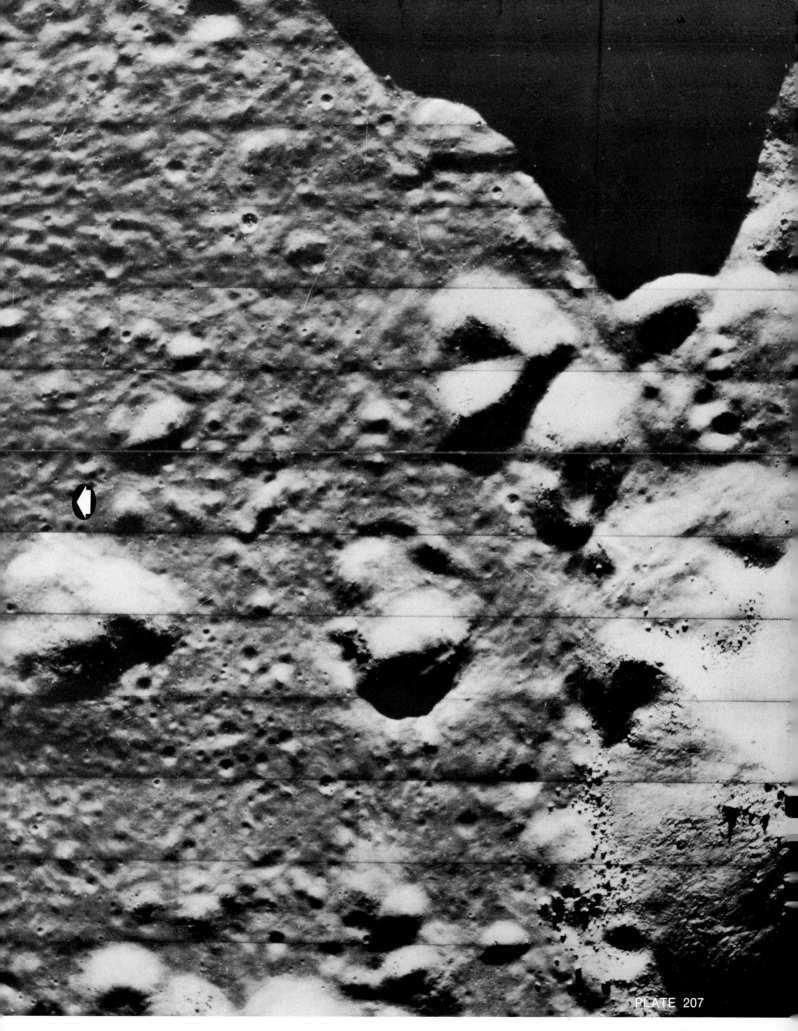

PLATE 207

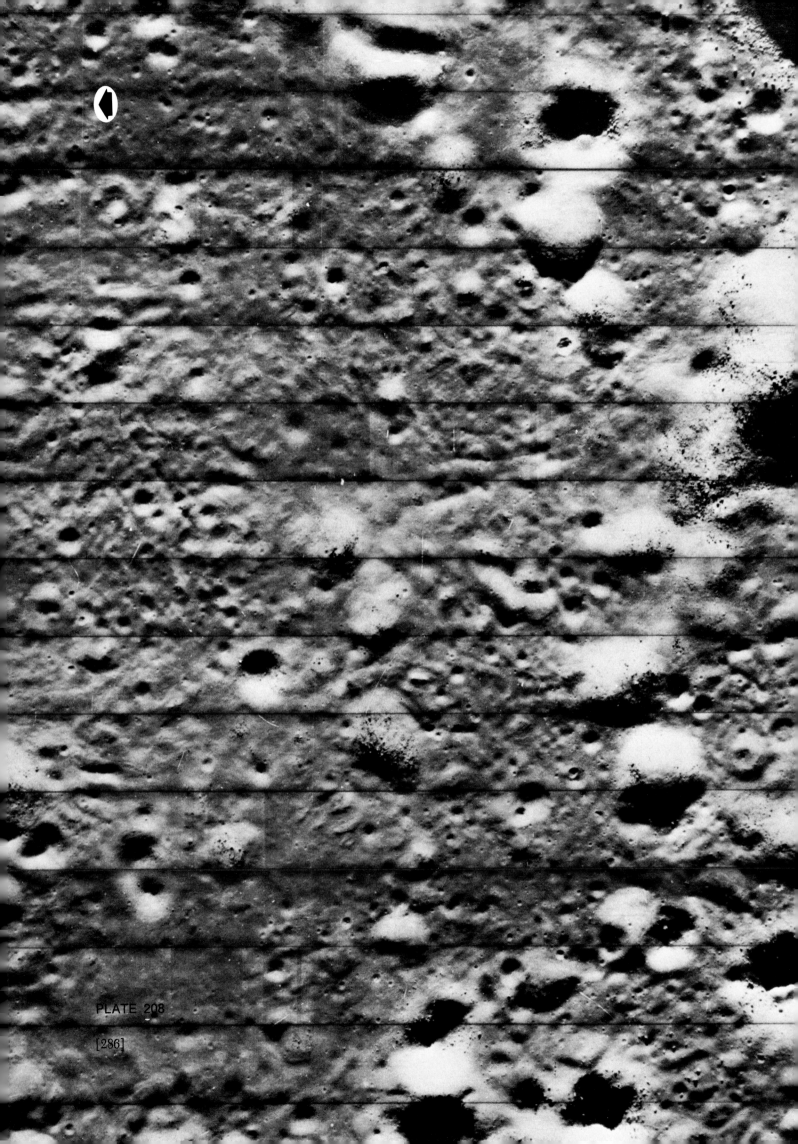

PLATE 208

[286]

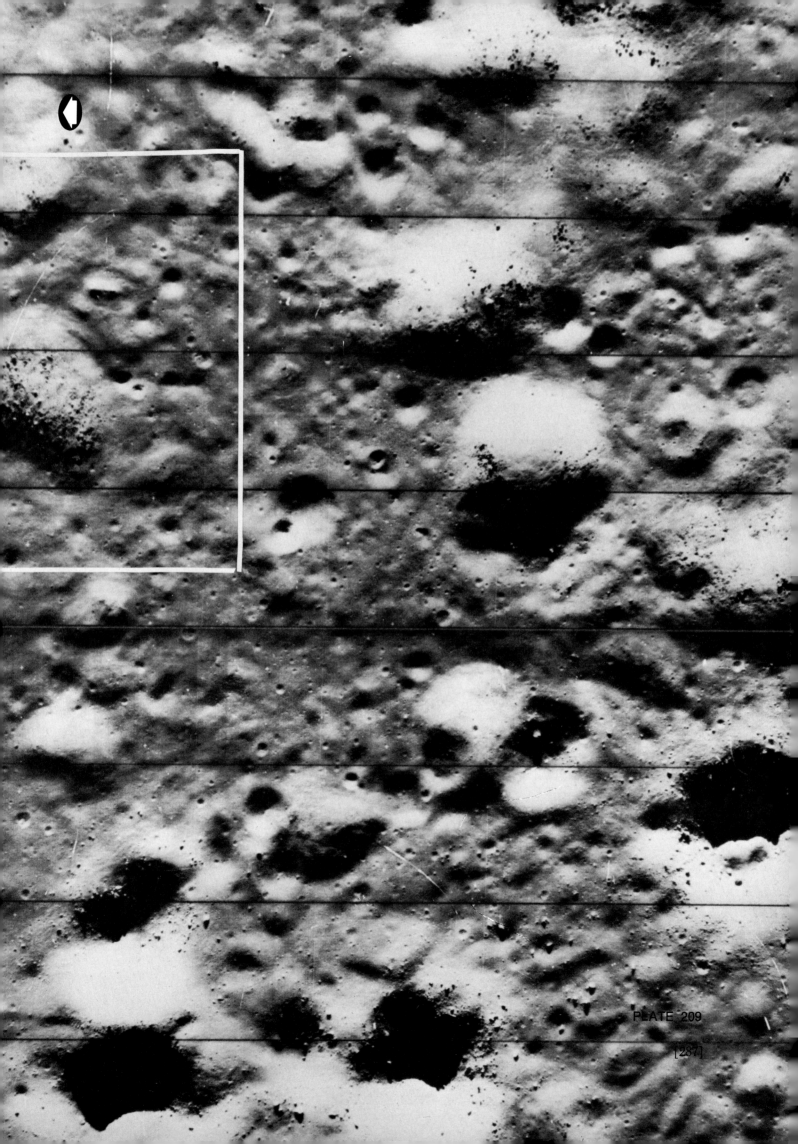

PLATE 209

PLATE 210

[288]

211. "Megalithic rings" of lunar boulders, 1–10 m in size, in the plains of Mare Tranquillitatis ($\lambda = 34°42$ E, $\beta = 2°81$ N) as photographed by Lunar Orbiter 2 on November 7, 1966 from an altitude of 45 km. The size of the field recorded on the photograph is approximately 450×360 m in size; framelet width is 215 m, and ground resolving power is 1–2 m. (Reproduced by courtesy of NASA-Langley and of the Boeing Company)

212. A field of small boulders in the plains of the Oceanus Procellarum close to the landing place of Surveyor 1 ($\lambda = 43°21$ W, $\beta = 2°45$ S) televised to the earth on June 13, 1966. The boulders are approximately a yard or so in size and surround a shallow crater-like depression. (Reproduced by courtesy of NASA and of JPL)

213. Men on the Moon. This photograph of eerie beauty was taken at Tranquility Base on July 20, 1969 by Neil Armstrong, and shows Edwin Aldrin standing in the magnificent desolation of the lunar landscape. Reflected in his visor is photographer Armstrong with his camera, the U.S. flag, the television camera and, at the right, a part of the "Eagle" Excursion Module. (Reproduced by courtesy of NASA-Houston)

214. Man's footstep in the lunar soil. (Reproduced by courtesy of NASA-Houston)

PLATE 211

PLATE 212

[291]

PLATE 213

PLATE 214

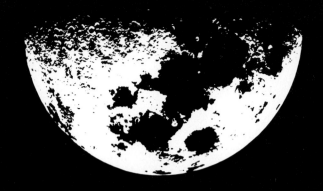

III. APPENDICES

Tables

TABLE 1. LIST OF LUNAR SPACECRAFT (1959–1969)

Name of Spacecraft	Origin	Date of Launching	Weight (in kg)*
		(a) Fly-by	
Luna 1	USSR	1959-January 2	362
Luna 3	USSR	1959-October 4	435
Ranger 3	USA	1962-January 26	330
Ranger 5	USA	1962-October 18	342
Luna 4	USSR	1963-April 2	1422
Luna 6	USSR	1965-June 8	1442
Zond 3	USSR	1965-July 18	960
		(b) Hard landing	
Luna 2	USSR	1959-September 12	390
Ranger 4	USA	1962-April 23	331
Ranger 6	USA	1964-January 30	365
Ranger 7	USA	1964-July 28	366
Ranger 8	USA	1965-February 17	367
Ranger 9	USA	1965-March 21	367
Luna 5	USSR	1965-May 9	1476
Luna 7	USSR	1965-October 4	1506
Luna 8	USSR	1965-December 3	1552
Surveyor 2	USA	1966-September 20	990
Surveyor 4	USA	1967-July 14	1038
		(c) Soft landing	
Luna 9	USSR	1966-January 31	1583 (100)
Surveyor 1	USA	1966-May 30	990 (292)
Luna 13	USSR	1966-December 21	1580 (100?)
Surveyor 3	USA	1967-April 17	1040 (302)
Surveyor 5	USA	1967-September 8	1006 (303)
Surveyor 6	USA	1967-November 7	1008 (303)
Surveyor 7	USA	1968-January 7	1010 (305)

(continued on next page)

---

* The weights in parentheses given for the soft-landers refer to those of the instrumented packages actually deposited on the lunar surface.

TABLE 1. LIST OF LUNAR SPACECRAFT (1959–1969), (*continued*)

Name of Spacecraft	Origin	Date of Launching	Weight (in kg)*
		(d) Orbiting	
Luna 10	USSR	1966-March 31	245
Orbiter 1	USA	1966-August 10	387
Luna 11	USSR	1966-August 24	1640
Luna 12	USSR	1966-October 22	?
Orbiter 2	USA	1966-November 6	390
Orbiter 3	USA	1967-February 5	386
Orbiter 4	USA	1967-May 4	390
Explorer 35	USA	1967-July 19	104
Orbiter 5	USA	1967-August 1	390
Luna 14	USSR	1968-April 7	?
Luna 15	USSR	1969-July 13	?
		(e) Re-entering	
Zond 5	USSR	1968-September 15–21	Unmanned
Zond 6	USSR	1968-November 10–17	Unmanned
Apollo 8	USA	1968-December 21–27	Manned
Apollo 10	USA	1969-May 18–26	Manned
Apollo 11	USA	1969-July 16–24	Manned**
Zond 7	USSR	1969-August 9–15	Unmanned
Apollo 12	USA	1969-November 17–24	Manned**

** Manned landings on the lunar surface.

TABLE 2. PLACE AND TIME OF UNMANNED SPACECRAFT LANDING ON THE MOON *

Spacecraft	Place of Impact (Longitude)	(Latitude)	Date and Time of Impact**	
		(a) Hard-landers		
Luna 2	0°.0	29°.1 N	1959-September 13	$22^h02^m24^{sec}$†
Ranger 6	21.52 E	9.33 N	1964-February 2	9 24 33.1
Ranger 7	20.58 W	10.63 S	1964-July 31	13 25 49
Ranger 8	24.65 E	2.67 N	1965-February 20	9 57 36.8
Ranger 9	2.37 W	12.83 S	1965-March 24	14 08 20
		(b) Soft-landers		
Luna 9	64°.37 W	7°.08 N	1966-February 3	$18^h44^m52^{sec}$
Surveyor 1	43.21 W	2.45 S	1966-June 2	6 17 37
Luna 13	62.05 W	18.87 N	1966-December 24	18 01
Surveyor 3	23.34 W	2.94 S	1967-April 20	0 04 53
Surveyor 5	23.18 E	1.41 N	1967-September 11	0 46 44.3
Surveyor 6	1.37 W	0.46 N	1967-November 10	1 01 5.5
Surveyor 7	11.41 W	41.01 S	1968-January 10	1 05 30

---

* Only those spacecraft are included which furnished lunar scientific information.
** Universal Time as observed on the Earth (not corrected for transit-time of the signals).
† The last stage of the carrier rocket of Luna 2 (1,121 kg in weight) impacted on the Moon 30 min later.

TABLE 3. KINEMATIC CHARACTERISTICS OF THE ARTIFICIAL LUNAR SATELLITES (1966–1969)

Spacecraft	Period	Inclination*	Altitude (in km)**		Injection into lunar orbit	End of mission	No. of days in orbit	Total no. of revolutions
			(Periselenium)	(Aposelenium)				
Luna 10	178min3sec	71°.9	349	1017	1966-Apr 3	1966-May 30†	67	
Orbiter 1	208.6	12.0	56	1853	1966-Aug 14	1966-Oct 29	76	547
Luna 11	178	27	159	1200	1966-Aug 18	1966-Oct 1†	34	
Luna 12	205	4	105	1740	1966-Oct 26			
Orbiter 2	208.4	11.9	49.7	1853	1966-Nov 10	1967-Oct 11	335	2289
Orbiter 3	208.6	20.9	54.9	1847	1967-Feb 8	1967-Oct 9	243	1843
Orbiter 4	721	85.5	2706	6114	1967-May 8	1967-Oct 6	70	225
Explorer 35	684	11.2	763	7670	1967-July 19	in orbit		
Orbiter 5	510.5	85.0	195	6029	1967-Aug 5		179	
"	503.5	84.6	100	6066	"		179	
"	191.3	84.8	99	1499	"	1968-Jan 29	179	1201
Luna 14	160	42.0	159	871	1968-Apr 10	?	?	?
Luna 15					1969-July 16	1969-July 21	5	?

* Inclination to lunar equator.
** Altitude above the mean lunar sphere of radius 1,738 km.
† Date of loss of radio-contact.

Bibliographical Notes

As mentioned earlier in the text, the greatest single source of data for this atlas has been the 43-inch reflector and 24-inch refractor at the Observatoire du Pic-du-Midi, in the hands of the University of Manchester observers supported by the U.S. Air Force, which provided more than one-third of all original photographs reproduced in this atlas. A more detailed description of the instruments as well as the techniques used in this work can be found, for example, in the writer's article, "Lunar Photography at Pic-du-Midi" in *Sky and Telescope, 33,* No. 4 (April 1967).

For a comprehensive report on the methods and results of space-borne photography of the Moon supplied by the Rangers, Surveyors, Orbiters and Apollos in the last decade, see a summarizing article on "Recent Observations of the Moon by Spacecraft" by L. D. Jaffe, in *Space Science Reviews, 9,* 491–609, 1969, containing many excellent reproductions of lunar photographs taken from space.

The reader who may wish to pursue more deeply various aspects of the subject matter included in the introduction to the atlas is referred to a recent more comprehensive treatise, *The Moon* by Z. Kopal (525 pp.), published in 1969 by the D. Reidel Publishing Company in Dordrecht, Holland.

Glossary

Albedo a technical term (of Arabic origin) customarily used to denote the ratio of the light incident from the Sun which is reflected by the surface of the Moon (or of a planet) in all directions.

Anomalistic Month a time interval of 27 days, 13 hours, 18 min and 37 sec, in which the Moon returns to the same place in its relative orbit around the Earth. It is somewhat longer than the "sidereal month" (after which the Moon returns to the same place in the sky) but shorter than the "synodic month" (in which the Moon returns to the same phase).

Apsidal line a line in space connecting two points of a relative elliptical orbit at which the revolving body comes closest to, and farthest apart from, the attracting center.

Auroral Zone region surrounding the magnetic poles of a celestial body, in which charged particles (of solar origin) are made to spiral down through the atmosphere and produce the "aurorae" (or polar lights).

Brecciated Lithosphere a stony shell surrounding the lunar globe, which has been fragmented into pieces by meteoritic impacts on the Moon.

Bright Rays splash patterns on the Moon surrounding lunar craters of (presumably) more recent origin. The difference in brightness between the ray material and the underlying ground may be due to an actual difference in composition between material thrown out by primary impacts from greater subsurface depth, as well as to a different degree of "bleaching" this material received during exposure to external influences on the lunar surface.

Circular Maria circular basins, bordered (partly or wholly) by surrounding mountain chains, characterized by flat floors filled with dark material. The presumption is strong that the circular maria represent nothing but large impact craters on the Moon, differing from numerous smaller formations of this type in size rather than in kind.

Clefts formations on the lunar surface caused by a sideslip (or subsidence) of ground along lines usually 50 to 100 km in length.

Cometary Heads and Tails the head of a comet consists typically of a "nucleus" which is nothing more than a cosmic iceberg of frozen hydrocarbons, surrounded by a semitransparent "coma" of evaporating gases. The "tail" of a comet represents the duct of escape of these gases (modulated by the local magnetic fields) and of their dispersal into interplanetary space.

Continents, Lunar as distinct from the maria—represent the prevalent type of lunar ground of higher reflectivity, covering more than 60 per cent of the Moon's visible face, and over 95 per cent of its far side. They constitute the oldest type of solid ground known to us in the solar system (about 4.6 billion years old); and their stony sculpture records traces of events much older than anything known to us on our own planet.

Cube-Corner Reflector an optical reflector deposited on the Moon by Apollo 11 and Luna 17, consisting of sets of three reflecting surfaces strictly perpendicular to each other and intersecting at a corner. The aim of this device is to reflect (by triple reflection) any light—such as the laser signals beamed upon it from the Earth—exactly backward no matter what the incident direction may have been.

Diurnal Libration an apparent motion of the Moon in the sky, because the terrestrial observer resides on the surface of a rotating planet. As a result, the Moon is bound to be nearer to us when it stands high in the sky than when it rises or sets; and the direction from which we see the face of the Moon varies slightly in the course of each night.

Domes formations of the lunar surface reminiscent of low bulging hills, often exhibiting a central depression on the top. Such formations are typically a few kilometers in size, and their altitudes are of the order of a few hundred meters. It is not known yet whether they possess any exact terrestrial analogues.

Draconic Month a time interval between two successive passages of the moon through the nodes of its relative orbit around the Earth. It is equal to 27 days, 5 hours, 5 minutes and 36 seconds.

Ecliptic the plane in which the Earth revolves around the Sun.

Evection a periodic perturbation of lunar motion, caused by solar attraction on the eccentricity of the lunar orbit. Its effects are apparent to the naked eye (as it can displace the position of the Moon by more than five times its apparent semidiameter) and was discovered in antiquity by Hipparchos.

Exosphere the topmost layer of a planetary atmosphere, where the mean-free-path of constituent gases becomes so long that collisions become unimportant and most molecules describe "ballistic trajectories" in the prevailing field of force.

Fluorescent Radiation radiation produced if the absorption of light by a given atom (or a molecule) is reemitted in successive installments (by a cascade process) to emerge as less energetic light of longer wavelength.

Gendarmes an Alpine term describing a particular type of ice formations (reminiscent of "men standing at attention") eroded by wind. Certain types of stony formations encountered on the Moon bear a rather striking resemblance to the icy "gendarmes" in the Alps.

Gravitational Constant a constant of proportionality $G = 6.668 \times 10^{-8}$ cm^3/g sec^2 in the relation between the magnitude of a mass, and the force of attraction exerted by it.

[304]

Hedgehogs on the Moon a playful term recently applied to certain types of low hills covered with boulders, from which the filler material has gradually been removed by repeated landslides.

Libration of the Moon in latitude and longitude—a small oscillatory motion of the apparent position of the Moon in the sky, because the relative motion of the Moon around the Earth is eccentric (giving rise to an "optical libration" of the Moon in longitude), and that its plane is inclined to that of the ecliptic ("optical libration in latitude"). Both these "optical" librations are a result of the particular properties of the motion of the Moon in space. In addition, the Moon exhibits also "physical libration" due to periodic motion of lunar globe around its center of gravity; but whereas the "optical" librations of the Moon may amount to several degrees selenocentrically (that is, as seen from the center of the Moon), the "physical" librations amount to scarcely more than 2 min of arc.

Light Curve of the Moon a graphical representation of the way in which the total light of the waxing or waning Moon varies with the phase in the course of a month.

Limb Darkening a diminution of the intensity of illumination of the disk of a spherical celestial body from its center toward the edge (a phenomenon which can be caused by absorption or scattering of light in its atmosphere).

Litho-Exhumation a geological process for sorting out stony debris according to their size by repeated shake-ups, which tend preferentially to lift coarser components of the mixture on top of the finer debris in the course of time.

Magma a generic term for rocks melted by internal heat (or any other process).

Magnetic Storms transient events caused by interaction of solar wind gusts with the magnetic field of the Earth (or the respective celestial body).

Maria ("seas") a distinct type of lunar terrain, covering some 37 per cent of the surface of the visible hemisphere of the Moon (though only some 5 per cent of its far side), characterized by low reflectivity (albedo) and a lesser degree of cratering than the continental areas. Cosmologically, the maria are probably younger than the continents.

Mascons mass concentrations, sensed by the motion of the lunar orbiting satellites to exist at a shallow depth below the surface of several "circular maria" on the Moon, and planted there probably by external impacts which gave rise to these formations.

Opposition Effect a surge in the intensity of moonlight near full Moon.

Periselenium sometimes referred to (in engineering literature) as perilune, denotes a point of the relative elliptical orbit at which the revolving body—such as a lunar artificial satellite—come closest to the center of the Moon. Half a revolution later, the same body happens to find itself at the greatest distance from the center, that is, in "aposelenium."

Perturbations deviations of the actual motion of the Moon in space from elliptical relative orbit, caused by forces other than the terrestrial attraction. If the disturbing force fluctuates periodically, it generally gives rise to "periodic" perturbations whose period may (but need not) coincide with that of the disturbing force. On the other hand, the perturbations which grow linearly with the time are called "secular"; the most conspicuous examples of these are the secular advance of the apsidal time of the lunar orbit, or the recession of the nodes.

Polar Aurorae   northern lights arising in regions around the magnetic poles of the Earth (or of any other planetary body) by an interaction of gas particles in the upper atmosphere with the "solar wind" particles spiraling down the magnetic lines of force.

Quiet Sun   the Sun around the time of the minimum of its eleven-year cycle of activity (or, at other times, in the absence of any local eruptions on its face).

Radar Echoes   reflections of radar signals sent out from the Earth and registered by the terrestrial receivers after a time lapse of 2.56 sec (which light takes to traverse the 768,000 km round trip of the mean Earth-Moon distance).

Regolith   a geological term to describe the fragmented stony layer (fractured by external impact) surrounding a planetary body with no atmosphere or liquid on its surface.

Rilles   characteristic formations in the marial regions of the lunar surface, which represent probably subsidence trenches in regions where the lunar regolith happens to be deep.

Seismic Waves   elastic waves in the solids produced by sudden (seismic) disturbances, such as can be excited by moonquakes, local subsidence, or meteoritic impacts.

Sidereal Month   a time interval which elapses between successive returns of the Moon to the same place in the sky, and equal to 27 days, 7 hours, 43 min and 12 sec.

Specular Reflection   reflection of light (or of other types of electromagnetic radiation) from smooth surface, following the laws of geometrical optics, as distinct from "diffuse reflection" in which the directions of the reflected waves become arbitrarily redistributed by surface irregularities.

Stereogrammetric Methods   of lunar research are concerned with studies of the three-dimensional shape of the Moon or its surface on the basis of stereoscopic comparison of photographs (or other data of the same area) secured from different locations in space.

Synodic Month   an interval of time which elapses between two successive identical phases of the Moon; it is equal to 29 days, 12 hours, 44 minutes and 3 seconds.

Syzygy   a line joining the positions of the "full" and "new" Moon.

White Light   integrated light of all colors emitted by a given celestial body.

Wrinkle Ridges   characteristic formations in the lunar maria, akin to the dunes in the desert, obviously of internal origin and possibly connected with solidified lava flows at the time when the maria were formed.

Index

[308]

NATIONAL AERONAUTICS AND SPACE ADMINISTRATION
LUNAR CHART

PREPARED BY THE AERO-
NAUTICAL CHART AND IN-
FORMATION CENTER,
UNITED STATES AIR
FORCE, UNDER THE DI-
RECTION OF THE DEPART-
MENT OF DEFENSE.

NOTES

The Lunar Surface Features shown on this chart are interpreted from the photographic records of Lunar Orbiter Missions I, II, III, IV, and V. Ray patterns and Albedo differences on the earthside limb and farside hemisphere are incomplete due to limitations of the source photographs. Horizontal position of the features is based on the ACIC Positional Reference System, 1969. Feature names are adopted from the 1935 International Astronomical Union Nomenclature System, as amended by I.A.U. in 1961 and 1964. The application of names is restricted to only those associated features that have been positively identified.

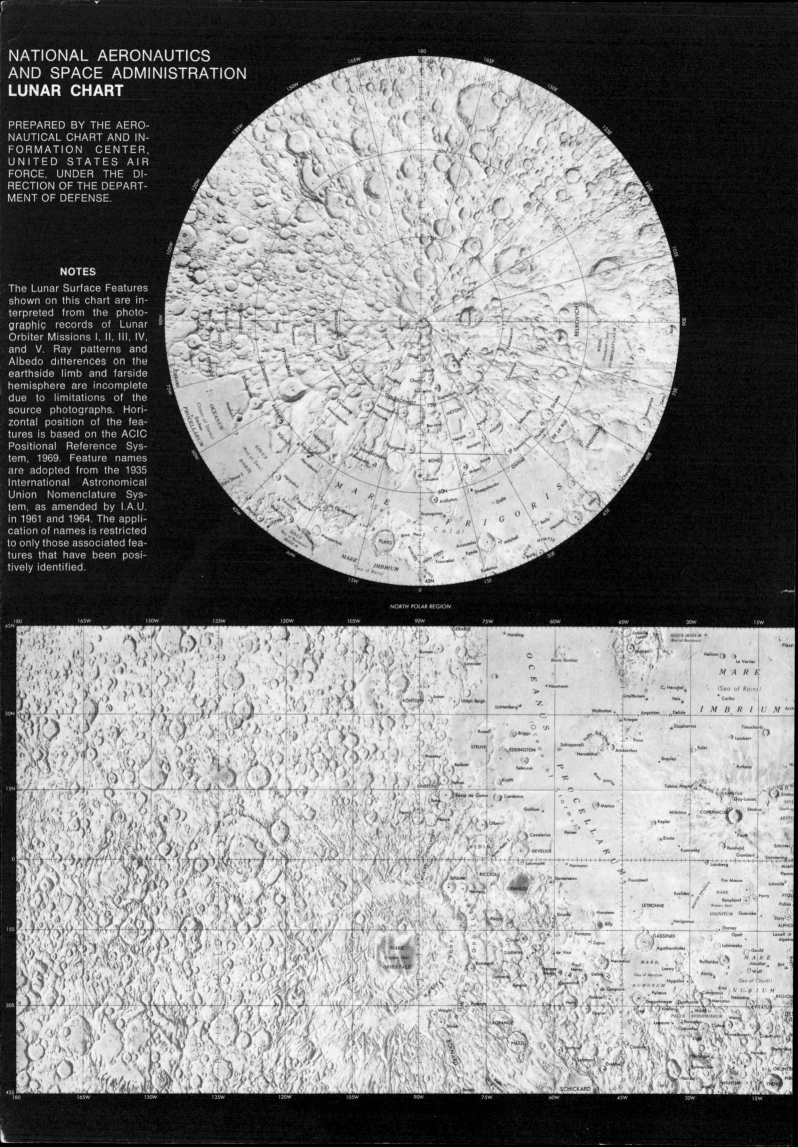

NORTH POLAR REGION